Slow Tech and ICT

Norberto Patrignani • Diane Whitehouse

Slow Tech and ICT

A Responsible, Sustainable and Ethical Approach

Norberto Patrignani
Politecnico of Torino and Università
Cattolica del Sacro Cuore
Milano, Italy

Diane Whitehouse
The Castlegate Consultancy
Malton, United Kingdom

ISBN 978-3-319-68943-2   ISBN 978-3-319-68944-9   (eBook)
https://doi.org/10.1007/978-3-319-68944-9

Library of Congress Control Number: 2017957090

Cover illustration: Pattern adapted from an Indian cotton print produced in the 19th century

Printed on acid-free paper

This Palgrave Macmillan imprint is published by Springer Nature
The registered company is Springer International Publishing AG
The registered company address is: Gewerbestrasse 11, 6330 Cham, Switzerland

Slow Tech and ICT—Endorsements

"It is wonderful to see the ecology of cyberspace taken up vigorously by people interested in promoting a good, clean and fair ICT. As a system engineer, I know our job in developing complex and integrated solutions is to put wisdom and long-term thinking into practice. This book is inspiring in its ambition to show us further perspectives to make this possible."

—Gino Romiti, *Innovation Director, Loccioni Group, Ancona, Italy*

"The book presents a challenging vision of how ICT should evolve in order to make a substantial contribution to the establishment of a sustainable, fair and liveable world. The authors show that it is crucial to escape from the present development model where natural resources are assumed to be limitless, time is an annoying constraint, and ethics is not an issue. This is an impressive book that makes a fundamental contribution by enriching the concept of sustainable development."

—Federico Butera, *Professor Emeritus of Environmental Physics, Politecnico of Milano, Italy*

"This book is to be highly recommended. It is an important contribution to the study of the relationship between sustainability and ICT in the long term. The authors argue for the introduction of Slow Tech as a new paradigm in the appropriate design of ICT and ICT systems. As a result, they suggest new policies for organisations and companies especially those companies aware of their social responsibility. This book is required reading in order to prepare the next generation not only of researchers and computer professionals, but also users, ICT designers, ICT industrialists and policy makers."

—Emilio D'Orazio, *Director of Politeia – Center for Ethics and Politics, Milano, Italy*

"Multiple moments and developments over the past decade mark out new directions towards designing ICT – creative and critical approaches desperately needed for the sake of good lives and for flourishing in today's technological era. The authors, leaders in these developments, authoritatively draw together insights and resources, some of which have been decades in the making. The result is a watershed contribution to the contemporary efflorescence of urgently needed collaboration between normative disciplines, including applied ethics and ICT design."

—Charles Ess, *Professor in Media Studies – Department of Media and Communication, University of Oslo, Norway*

"A wonderful and well-researched book that will greatly benefit the education of technology students and leaders worldwide. The concepts presented in the book provide a challenge to our thinking about Information and Computer Technology and our societal appetite for consumption of technological innovations. I look forward to incorporating it as a tool in my classroom."

—Rebecca Lee Hammons, *Associate Professor – Center for Information and Communication Science, Ball State University, USA*

"In the midst of what feels like unrelenting increases in the pace of human activities and technological change, this book offers a simple but powerful alternative idea, Slow Tech. The authors weave a tapestry around the idea putting it in historical context and providing an optimistic vision for our potential future."

—Deborah G. Johnson, *Olsson Professor of Applied Ethics, Emeritus, STS Program, Department of Engineering & Society, University of Virginia, USA*

"'ICT will very soon solve almost all our problems and it will satisfy all our needs.' At least that is how we will feel. Then 'we' will not be needed anymore. Our thinking, our making of choices, our responsibility, and our anxiety will all be gone. But, is this what we really want? If yes, then this will be heaven. If not, we have to do something about it and, in order to have a chance, we need to prepare ourselves. We need a clear mind and practical tools to use. Slow Tech offers some of these."

—Iordanis Kavathatzopoulos, *Professor, Department of Information Technology – Division of Visual Information and Interaction, Uppsala University, Sweden*

"This book offers a roadmap for ICT based on the 'slow perspective'. It is this slow perspective which is the compass to navigate new directions for a more responsible, sustainable, and ethical approach to ICT. A truly refreshing perspective for imagining the future we want and need."

—Viola Schiaffonati, *Associate Professor of Computer Ethics, Politecnico di Milano, Italy*

"This book makes an important contribution to the question of how best to balance benefits and disadvantages. By drawing on the established idea of Slow Food, the authors develop a vision of future technology development that emphasises the importance of taking time to reflect, which is required to promote responsibility, sustainability and ethics. It will give (slow) food for thought to researchers, developers, the ICT industry and policy makers, and thereby contribute to ensuring that future ICT benefits individuals and society."
—Bernd Carsten Stahl, *Professor of Critical Research in Technology, Centre for Computing and Social Responsibility, Faculty of Technology, De Montfort University, United Kingdom*

"Inserting the word 'time' into the reasoning about technologies is precisely what was missing for both present, and future, generations to build up an ethical approach to ICT. This book flows swiftly. It provides a large, clear range of reference thinkers and virtuous examples. It offers its readers a coherent set of materials, to enable them to reason together with the authors. In order to develop the means for a positive Anthropocene balance with the ecosystem, I personally hope that ethical human beings will greet this Slow Tech evolution."
—Clelia Caldesi Valeri, *PhD Cultural Heritage and Infosphere, Researcher in the ethics of culture and ICT, Italy*

FOREWORD

In the first decade of the twenty-first century, time is in psychosocial shortage. This is despite the fact that, in today's net era, people have access to technology that can offer them more time than ever and free them from paid work—both the repetitive and physically heavy work that dominated during periods of industrialism and also more and more mental work. Why did this wonderful technology, called information and communication technology (ICT), not provide people with lots of free time to enjoy life, increase the quality of life, and care about nature—comparable to the Greek golden era?

The main title of this book, *Slow Tech and ICT*, is indeed provocative and hits the nail on the head. Why go for slow technology when speed is what is wanted from machines? What happens when the purpose of all research and development is to develop faster, smaller, cheaper products and services through the interaction and convergence of permanently new components? The authors undertake an impressive analysis of this challenge. They also point out a potential direction to take and develop a new roadmap to achieve desirable societal and human goals.

The keywords, *good*, *clean*, and *fair*, which are directly linked with Slow Tech, are pearls in a chain that we all should wear around our necks. These Slow Tech guidelines are applicable to all countries. They involve values that we all have to care about, independent of where we are located and at what stage we are in the digital transformation.

Just by taking a quick look at the book's Preface, we are straightaway curious to learn more about the Slow Tech philosophy.

Royal Institute of Technology   Gunilla Bradley,
Stockholm, Sweden
Summer 02017

Note on Dates

Throughout this book five figures are used instead of four when giving dates. For example, 2017 is written as 02017. This follows the convention adopted by the Long Now Foundation in response to the problem of the so-called millennium bug. More about this can be read in Chap. 4.

PREFACE

Dear Readers,

Information and communication technology (ICT) is disrupting people's lives. In an interconnected world made up of a network of nodes, every individual is becoming a node in such a network.

Based on our own working lives and experiences in the field of ICT, we now have the impression of having reached a turning point. We feel this is absolutely an appropriate time to reflect on the design and use of a human-centred ICT. With this aim in mind we want to share with you our need to find a new direction, a roadmap—a new route or sea lane for ICT—which is not pre-defined, but which people need to develop together. We ourselves do not know the precise way forward nor do we hold the truth in our hands. Instead, we propose what we hope are some useful starting points to use as foundations when searching for a route to the future.

To find this new route, we need a compass. As a possible compass, we put forward the concept of *Slow Tech*: it promises everyone the opportunity to travel in interesting new directions—particularly ones that offer a responsible, sustainable, and ethical approach to ICT.

Who do we want to involve, and why, in this search for a suitable approach and appropriate uses of ICT? We want to involve all the stakeholders in the ICT value chain and in what we call the ICT stakeholders' network. We intend to include in this discussion ICT designers (current designers, the next generation of designers, and young engineers), ICT professionals, ordinary people at work and at home—people at large, policy makers, and personnel in corporations and companies—as well as all the different types of organisations involved with ICT.

From our own personal and professional experiences, we remember some of the main reservations that ICT designers and developers expressed already 30 years ago about the exciting participatory and collaborative methods that were then being introduced: '... We never get the time to use the methods. ... Our employers would never allow us to use these tools ... The activities do not fit with the ethos of the corporations for which we work', said colleagues at the time. Unfortunately, such objections are still heard today.

In the year that we call 02017, there is, however, a completely different scenario playing out from the one that was current in 01987. Today, people are more alert to the possibilities offered by many technological and societal developments. In general, people are more educated about the uses of technologies; they recognise more clearly the level of disruption caused by ICT in their lives—even if, for younger generations, it can be difficult to find a balance between their online and offline lives. Researchers and scientists are investigating the environmental impact of ICT. Corporate ethics are undergoing more profound scrutiny from society, and ever larger numbers of organisations are conscious of the importance of developing strategies on corporate social responsibility: they are more and more aware that a sustainable business—from an economic point of view—needs a sustainable strategy from a social and environmental viewpoint. Last but not least, if three decades ago there did not seem to be enough time to take on these grand challenges, the pressures are considerably more present in 02017. This is not only the case in the industrialised world but also in low- and middle-income countries where ICT is being adopted rapidly. It is this growing tension that we are seeking to overcome with *Slow Tech and ICT*.

We want to introduce the notion of *Slow Tech and ICT* by offering you a sense of what the past has held in relation to ICT and what the future can hold, and by describing a useful set of stories and anecdotes enhanced by illustrations. We sincerely hope that this book can help people to develop new guidelines for the appropriate design and use of ICT and, as a result, suggest new policies for organisations and companies.

We are happy to join you on this enriching journey towards a more responsible, sustainable, and ethical approach to ICT.

Milano, Italy   Norberto Patrignani
Malton, UK   Diane Whitehouse
Summer 02017

ACKNOWLEDGEMENTS

The first co-written words in this book's manuscript date from 02011. They form just the start of a journey that we hope we will continue fruitfully for some time to come.

Some of the events and activities that have coloured the developments of our ideas have been: the Brain, Body, Being Conference (Maribor, Slovenia, June 02010), a workshop on sustainability held at the Politeia centre (Milano, Italy, June 02011), and a walk around the city of Ivrea, Italy, where the headquarters of the historical computer company led by the 01950s visionary, Adriano Olivetti, is located (spring 02015).

A number of people have consciously influenced our writing and thinking, in one case going back at least several years. Among those whom we would like to thank for their inputs to our thinking are Gunilla Bradley, for her ideas on wisdom and making conscious decisions; Fiorella De Cindio, for her input on the historical foundation of e-democracy; Emilio D'Orazio, for his informed suggestions about social responsibility; Marc Griffiths, who highlighted the possible parallels between Slow Tech and futurism; Magda Hercheui, with whom there were several discussions in mid- and late 02011 about the growth of thinking about war and negativism which are sadly even more pertinent today; Deborah Johnson for a series of inspiring conversations, including an in-depth discussion on global differences on the terrace of a Parisian cafe; George Karageorgos, with whom there were similar discussions in mid-02011 on war and violence, especially the social and economic dilemmas facing Greece; Iordanis Kavathatzopoulos, for his suggestions relating to philosophy; Enrico Levati, for his kind introduction to the Slow Food movement; Bernadett

Koteles, with whom there were several intense conversations about the importance of positive thinking and acting; May Pettigrew, who back in 02006–02008 was already drawing our attention to the parallels of the beginning of this century with the start of the twentieth century; Mick Phythian, who suggested parallels between our own thinking and the 02012 work of Luciano Floridi; Magda Rosenmöller, who offered insights into how change takes time and how current developments often have their roots long in the past and who hence stimulated Diane Whitehouse and Petra Wilson towards their collaborative work on e-health visions of the future (Rosenmöller et al. 2014); and last, but certainly not least, Clelia Caldesi Valeri, for her gentle conversations.

In writing this book, we have benefitted from feedback offered by many individuals at several presentations that we have given over the years, particularly in the context of events held by the International Federation for Information Processing and ETHICOMP, among others. The attendees posing us questions or engaging us in pertinent conversations have not always been known to us and we were not always able to speak with them face to face. However, all of these interventions have had their significance. Even more applicable have been the thoughtful, often anonymous, reviews and commentaries on papers and articles that we have written together, particularly in the period since 02013.

As we emphasise in Chap. 2 of this volume, the work of Carlo Petrini, Alexander Langer, René von Schomberg, and Joseph Weizenbaum has exerted considerable influence on us. We call these four people pneumatophores—these are the individuals who have acted as our 'root structures', insofar as their ideas have given us oxygen to feed our own concerns and considerations.

BIBLIOGRAPHY

Rosenmöller, M., Whitehouse, D., & Wilson, P. (Eds.). (2014). *Managing eHealth: From Vision to Reality*. London: Palgrave Macmillan.

CONTENTS

Contents

Slow Tech List of Abbreviations

ACM	Association of Computing Machinery
BINAC	Binary Automatic Computer
BRICS	Brazil, Russia, India, China and South Africa
CPSR	Computer Professionals for Social Responsibility
EDVAC	Electronic Discrete Variable Automatic Computer
EGAIS	The Ethical GovernAnce of emergIng technologieS
ETICA	Ethical Issues of Emerging ICT Applications
FLOPS	Floating Point Operations per Second
fMRI	Functional Magnetic Resonance Imaging
IaaS	Infrastructure as a Service
IBM	International Business Machines
ICANN	International Corporation for Assigned Names and Numbers
ICT	Information and Communication Technology
ICT4S	ICT4Sustainability
IFIP	International Federation for Information Processing
IPCC	Intergovernmental Panel on Climate Change
ISO	International Organization for Standardization
IT	Information Technology
Kbit	Kilobit
NY	New York
OECD	Organisation for Economic Co-operation and Development
PaaS	Platform as a Service
PD	Participatory Design
PET	Positron Emission Tomography

RRI	Responsible Research and Innovation
SaaS	Software as a Service
SoDIS	Software Development Impact Statement
STS	Science, Technology and Society
TCP/IP	Transmission Control Protocol/Internet Protocol
UNIVAC 1	UNIVersal Automatic Computer I
WEEE	Waste Electronic and Electrical Equipment

LIST OF FIGURES

List of Figures

CHAPTER 1

Introduction

Abstract This chapter is an introduction to the whole book. It describes the origins of the authors' thinking about Slow Tech, stimulated by the comparison with the international Slow Food movement. Slow Food as an approach has been paralleled by several other trends in 'slow'. In the field of technology, it has also had different manifestations: it has been associated with mindfulness, appropriate mobile telephony, and the need for a growth in resilience. As authors, we see going 'slow' as a call for reflection on the kind of world, and technologies, that we want and need. We call for a collective dialogue and collaborative movement on making ICT design more responsible, sustainable, and ethical.

Keywords Conversation • Compass • Ethical • Responsible • Sustainable • Slow Tech

Conversations need time. Since we first discussed many of the challenges apparent in this book, we have—by meeting three or four times a year and working together at a distance—formulated a new way of thinking about information and communication technology (ICT) that we call Slow Tech.

This book fits well with many of the slow approaches which have been in development for a quarter of a century in various fields. Together, and with colleagues, we have drawn principally on parallels with the Slow

© The Author(s) 2018
N. Patrignani, D. Whitehouse, *Slow Tech and ICT*,
https://doi.org/10.1007/978-3-319-68944-9_1

Food movement (Petrini 2011) to develop this new approach to ICT: an approach that emphasises the importance of responsibility, sustainability, and ethics.

The Slow Food movement was launched in Northern Italy in the mid-01980s and has been spreading steadily internationally ever since. It has led to a more profound understanding of food, crops, and agriculture. According to Slow Food, food must be *good* (or taste good), *clean* (be produced following criteria that respect the environment), and *fair* (so as to ensure the rights of farmers). The slow concept has been adopted in many other fields too—in art, fashion, families, the home, medicine (see www.slowmedicine.it), and journalism (see www.slow-journalism.com).

The term Slow Tech has emerged in at least three other areas related to technology: a form of ICT typified by a concern for reflection, mental rest, and mindfulness (Hallnäs and Redström 2001), a description of the cultural and social impacts of mobile ICT devices (Kopomaa 2007), and an investigation of problems related to the lack of resilience of modern-day products (Price 2009).

We have our own way of interpreting the meaning of Slow Tech, chiefly in relation to Slow Food (Petrini 2011), although we have also drawn on the influences of a number of other technology-related thinkers and activists (see Chap. 2 on pneumatophores). Our starting point is that slow ideas can certainly be transferred to ICT.

Slow Tech provides people with 'the brakes required in the technological Indianapolis' (Patrignani and Whitehouse 2013: 389). It offers people more time for reflection and for the processes needed to design and use ICT that takes into account human well-being (*good* ICT), the whole life cycle of the materials, energy, and products used to create, manufacture, power, and dispose of ICT (*clean* ICT), and the working conditions of workers throughout the entire ICT supply chain (*fair* ICT).

As a result of this thinking, we propose that Slow Tech can act as a compass to help people to understand the directions needed for a new roadmap for computing. A Slow Tech movement could enable human beings to concentrate on aspects of their lives that are more important and meaningful than simple speed and efficiency. Hence, we encourage our book's readers to work collaboratively on finding new directions for ICT. This growing cooperation could mean the start of activities that affect positively and constructively people's individual, social, and corporate lives.

There are some signs that such a movement is starting already. In their home life, people are growing their own fruits and vegetables, recycling, knitting, and sewing: they are also turning off their machines and are instead reading books, writing letters, drawing, and starting conversations. In the commercial world, the demand for a new relationship between technology and society is also being recognised (Maketechhuman 2015).

* * *

This book acts step-by-step as a gentle, general introduction to the notion of Slow Tech in terms of ICT: an ICT that is *good* (human-centred), *clean* (environmentally sustainable), and *fair* (meaning that it respects the human rights of all the workers involved along the entire ICT value chain).

The chapters reflect the book's title with its focus on a responsible, sustainable, and ethical approach to ICT. Each chapter contains its own abstract and keywords.

The book should be read as an encouragement to move in a Slow Tech direction and to search out concrete ways and means of acting in a Slow Tech way on a day-to-day basis. It is intended that the volume offers plenty of examples and illustrations of people, companies, initiatives, and movements that can help to provide inspiration for this new way of thinking and behaving about ICT, and ensuring responsibility, sustainability, and ethics.

REFERENCES

Hallnäs, L., & Redström, J. (2001). Slow Technology; Designing for Reflection. *Journal of Personal and Ubiquitous Computing, 5*(3), 201–210.

Kopomaa, T. (2007). Affected by the Mobiles: Mobile Phone Culture, Text Messaging, and Digital Welfare Services. In R. Pertierra (Ed.), *The Social Construction and Usage of Communication Technologies: Asian and European Experiences* (pp. 48–59). Quezon City: University of the Philippines Press.

Maketechhuman. (2015). http://www.wired.com/category/maketechhuman. Accessed [9 October 2017].

Patrignani, N., & Whitehouse, D. (2013). Slow Tech: Towards Good, Clean, and Fair ICT. In T. W. Bynum, W. Fleischman, A. Gerdes, G. M. Nielsen & S. Rogerson (Eds.), *The Possibilities of Ethical ICT*, Proceedings of ETHICOMP 2013—International Conference on the Social and Ethical Impacts of Information and Communication Technology (pp. 384–390), Print & Sign University of Southern Denmark, Kolding.

Petrini, C. (2011). *Buono, Pulito e Giusto*. Torino: Einaudi.

Price, A. (2009). *Slow-Tech: Manifesto for an Over-Wound World*. London: Atlantic Books.

CHAPTER 2

Pneumatophores

Abstract This chapter introduces the four people whom the authors consider as foundational to their thinking—they call them pneumatophores. It is a recognition of the main thinkers and doers who have inspired this volume: Carlo Petrini, the founder of the Slow Food movement; Alexander Langer, the challenger of the Olympic motto with its focus on higher, faster, and stronger; René von Schomberg, the European civil servant who reoriented research and policy work towards responsible innovation; and Joseph Weizenbaum, the groundbreaking computer scientist who reminded his readers of the highly important need in the computing world for human choice and responsibility.

Keywords Change • Desirability • Heritage • Human judgement • Oxygen • Pneumatophores • Roots • Responsible innovation • Slow Food

In botanical studies, pneumatophores are—in marshes or bogs—the parts of tree roots that access the air and nurture other parts of trees that otherwise remain hidden. In human communities, pneumatophores are special people who act as spirit bearers and who inspire others.

There have been four main individuals whom we consider as Slow Tech pneumatophores. They are Carlo Petrini, Alexander Langer, René von

© The Author(s) 2018
N. Patrignani, D. Whitehouse, *Slow Tech and ICT*,
https://doi.org/10.1007/978-3-319-68944-9_2

Schomberg, and Joseph Weizenbaum. These are the four main individuals who have provided the oxygen that operates as the basis of our ideas.

Here we describe briefly the intellectual and practical heritage of each of these individuals, so that it is possible to understand in some depth the origins of our thinking. Their contributions—when considered as a whole—have provided an overall philosophy that can be regarded as a new ecology of mind (Bateson 1972): this is an approach that, on the one hand, offers the opportunity to have more time for thinking, observing, and choosing and, on the other hand, for people to enjoy their short lives more.

In this book we have applied this holistic philosophy specifically to ICT: the result is *Slow Tech and ICT*—a way forward for reaching good, clean, and fair ICT that is responsible, sustainable, and ethical.

2.1   CARLO PETRINI

One person, in particular, has focused on the notion of slow: he is Carlo Petrini.

Petrini was born in Bra in Piemonte, Italy. His mother was from a small farming community and his father was a railway employee. Petrini studied sociology at Trento University and began to write about food in 01977. He promoted the culture of conviviality and good food and wine. He coined the term 'eco-gastronomy' before founding, with colleagues, the Slow Food association in Bra on 9 December 01989.

Slow Food was begun by Petrini and a group of activists with the goal of 'countering the rise of fast-food and fast life'. Its foundation was at the origin of the, now much more internationally widespread, Slow Food movement which has gathered more than 100,000 members in 150 countries. Among the aims of this global, grassroots organisation are to link together the pleasure of good food with a commitment to the community and environment, counter the rise of a fast life, and focus on where the food eaten comes from and how the choices about eating it affect the rest of the globe (Slow Food 2011).

Petrini has written often about the relationship between food and the environment and sustainable development and material culture. He has received many awards: in 02003 he was given an honorary degree in Cultural Anthropology by the University Institute of Naples and in 02006 an honorary degree in Human Letters from the University of New Hampshire in the United States for being a 'revolutionary precursor and founder of the University of Gastronomic Sciences'. In 02004 he was nominated Hero of the Year by *TIME* Magazine Europe, and, in 02008, he was named 'one of the 50 people who could save the planet' by *The Guardian* newspaper.

One of the main volumes that describes the growth of the Slow Food movement is *Buono, Pulito e Giusto* (*Good, Clean, and Fair*) (Petrini 2011). In that book, Petrini describes the fundamental principles of Slow Food: food must be *good* (or taste good), it must be a pleasure to eat but, while eating, people should also reflect on and think about where the food comes from. Good food must be selected according to several properties. These include the attribute of quality. Yet, Slow Food must also be produced following the criteria that respect the environment (it must be *clean*). It must be based on local histories and traditions—such as ancient recipes based on the oral transmission of wisdom from one generation to the other. It must promote biodiversity, sustainability, and the rights of farmers (hence, it must be *fair*) (Petrini 2011).

The thinking, and the activities, of Petrini particularly inspired us as authors in terms of the main dimensions of Slow Tech: *good*, *clean*, and *fair* ICT. These three fundamental characteristics are of immense importance. To achieve them concretely, in terms of technology, poses considerable challenges both to people at large and the entire ICT industry. The size of that challenge convinced us to pick up the gauntlet ourselves, which is precisely why we started the enterprise of translating the three principles of Slow Food into Slow Tech.

2.2   ALEXANDER LANGER

Alexander Langer (1946–1995) was an Italian thinker. During the 01980s, he focused on the socially desirable side of change. The foundation dedicated to Langer's memory introduces him as having been concerned with conscious acts of self-restraint and having insight to the fact that an increasingly artificial world is no response to unjust, polluted, socially degraded, violent living conditions which are void of any purpose (Langer Foundation 2015).

Langer provided a major challenge to the wording of the Olympic motto: *Citius, Altius, Fortius*. This phrase, which draws together the three notions of faster, higher, and stronger, was coined by Henri Didon (1840–1900), a French Dominican educator, for a youth gathering in 01891. Three years later, in 01894, his friend, Pierre de Coubertin, proposed the same expression at the establishment of the International

Olympic Committee. At the 01924 games in Paris, it became the official motto of the Olympics. Indeed, in the adoption of this motto, it is possible to recognise a parallel with the birth of the Futurist movement of the 01900s (that celebrated 'the beauty of speed') (Marinetti 1909).

The simple three-part Olympic motto—with its focus on speed, height, and strength—has now pervaded all the small spaces, gaps, and interstices in human society. Speed and competition, in particular, are two of the myths of the contemporary era. A fixation with speed and competition at any cost means that people care less and less about human or planetary limits and cycles. When applied not just to sports but to other human activities, competition shifts people's attention onto very, very short-term goals. Today, however, people are starting to acknowledge that there are in fact some limits to speed and competition and that it would be far wiser to coexist with some rather longer-term cycles.

A more sustainable approach to living, from both social and environmental points of view, is now needed: an intensive rethinking of lifestyles according to longer-term perspectives. How can these principles be applied to people's lives (e.g. their health, their old age)? Can they be used to critique the basic assumptions that were followed throughout the twentieth century—of competition at any cost and the maximisation of short-term profit? Can they help to challenge developments that might otherwise be non-future proof and unsustainable?

In the twenty-first century, from the social and environmental points of view, societies have to face particularly tough, complex challenges. Recent ecological developments help to question the century-long paradigm of speed and achievement that many human beings have followed slavishly. The profound ecological, social, and economic crisis that humankind is now facing highlights the need to search instead for a new model that is more fitting for the future at all sorts of levels, including the individual, community, and societal. These contemporary approaches propose that only the introduction of a new model of societal organisation, a more socially and environmentally focused framework, will enable human survival in the long term. They enable a basic questioning: whether humankind, on behalf of planet Earth, should accept the extreme consequences of taking a Darwinian perspective to the treatment of 'the unfit' of the human species.

Langer already reflected deeply on these issues two decades ago (Langer 1996). He contrasted the tension that exists between the surrounding economic framework of competition and a new concept of well-being and well-living. He argued that an approach to life and living that is slower, deeper, sweeter needs to be adopted. Equally, a change in ecological direction is feasible solely when it is actually socially desirable. Langer's philosophy has much in common with what today one can call Slow Tech. His thinking inspires a need to reflect on the roots of social, economic, and environmental change. To take place, change must itself be desirable.

It is for this reason that in our proposal for the take-up of Slow Tech, we place such a focus on the concepts of slower, deeper, and sweeter ICT. In complete contrast to the Olympic motto of faster, higher, and stronger, we propose instead that there should be an ICT that is 'lentius, profundius, suavius' (slower, deeper, sweeter) (Langer 1996), a human-centred and planet-centred ICT and one which—taking the subtitle of this book—is responsible, sustainable, and ethical.

2.3   René von Schomberg

René von Schomberg continues to be one of the main thinkers at the frontline of European public policy, a philosopher who concentrates on the relationships between science, technology, and society. He has been working for the European Commission since 01998. One of his main contributions in this sphere of public policy development is in relation to ICT. His argument is that a new kind of innovation is needed. According to a 02011 report published by the European Commission, ICT and responsible innovation must be socially desirable and inclusive, environmentally sustainable, and ethically acceptable (von Schomberg 2011).

These three concepts of social desirability, environmental sustainability, and ethical acceptability in von Schomberg's work are developed according to two dimensions: the product dimension and the process dimension. The product dimension refers to product development and its characteristics. The process dimension refers to various normative anchor points identified by von Schomberg (2011), for example, technology assessment, foresight, risk assessment, and innovation governance, and the shift towards deliberative democracy, including the use of codes of conduct, standards, and certification schemes.

von Schomberg's contributions to contemporary policy thinking with regard to ICT contain many hooks to the general philosophy and background for the more proactive computer ethics that we call Slow Tech: designing, creating, and using an ICT that is socially desirable, environmentally sustainable, and ethically acceptable.

2.4   JOSEPH WEIZENBAUM

Joseph Weizenbaum (1923–2008) was a leading artificial intelligence researcher who had a profound personal and professional commitment to humanism. He began one of the first reflections on the different speeds of humans and computers in the twentieth century (Weizenbaum 1976). He was Professor Emeritus of Computer Science at the Massachusetts Institute of Technology and a thinker, writer, and activist.

Weizenbaum was the first recipient of the Namur Award in 01991 (the award was later given to further recipients on ten other occasions by the International Federation for Information Processing's Working Group 9.2 on the Social Accountability of Computing). Award recipients have all been renowned individuals who have made significant intellectual, academic, and political contributions to furthering the social implications of information technology at an international level. In his acceptance speech, Weizenbaum made it clear that making meaningful decisions takes time, and human decision-making has quite a different rhythm from the speed of operation implicit in ICT.

Weizenbaum reflected seriously on the limits of ICT by suggesting that human beings should never delegate important decisions to computers (Weizenbaum 1976). He argued that computers will always lack human qualities, such as compassion and wisdom, and they are programmed only to perform computational activities, that is, programmable decisions. Weizenbaum separated decisions from choices. Decisions can be a computational activity and can be programmed inside computers. In contrast, human beings are able to make choices, and choices are the products of judgement, not calculation. It is this capacity for choice that ultimately makes human beings human, and this human judgement includes emotions. Hence, Weizenbaum suggested that human functions that need judgement, respect, understanding, care, and love ought not to be substituted to be made by computers. ICT development is thus a matter of human choice and responsibility (Weizenbaum 1976).

When we think of Slow Tech as a compass for making future decisions on ICT, Weizenbaum has been for us one of the four main contributors towards this direction setting in a responsible, sustainable, and ethical approach to technologies.

References

Bateson, G. (1972). *Steps to an Ecology of Mind*. Chicago: Chicago University Press.

Langer, A. (1996). La conversione ecologica potrà affermarsi solo se apparirà socialmente desiderabile, Colloqui di Dobbiaco. In E. Rabini & A. Sofri (Eds.), *Il viaggiatore leggero. Scritti 1961–1995*. Palermo: Sellerio editore.

Langer Foundation. (2015). *The Alexander Langer Foundation*. http://www.alexanderlanger.org/en/668. Accessed [9 October 2017].

Marinetti, F. (1909, February 20). Le Futurisme. *Le Figaro*. Published also by several Italian newspapers as *Manifesto del Futurismo*.

Petrini, C. (2011). *Buono, Pulito e Giusto*. Torino: Einaudi.

Slow Food. (2011). http://slowfood.it/international. Accessed [9 October 2017].

von Schomberg, R. (Ed.). (2011). *Towards Responsible Research and Innovation in the Information and Communication Technologies and Security Technologies Fields*. European Commission, Directorate General for Research and Innovation, Luxembourg, Publications Office of the European Union. http://ec.europa.eu/research/science-society/document_library/pdf_06/mep-rapport-2011_en.pdf. Accessed [9 October 2017].

Weizenbaum, J. (1976). *Computer Power and Human Reason: From Judgment to Calculation*. New York/San Francisco: W.H. Freeman.

CHAPTER 3

Speed Limits in Cyberspace

Abstract This chapter reflects on technology and its relationship with time and human limits, including whether there are any limits in cyberspace. ICT is now developing very fast. It is modifying both human beings' ways of doing things as well as who they are as people. Similarly, how human beings perceive time has also recently been changing rapidly. People are undertaking many more tasks in swift succession and in parallel. People find these changes in behaviour highly demanding. These rapid changes can actually act as an encouragement to think about the very opposite of speed. A set of Slow Tech guidelines can help human beings to live at a pace that is more responsive to, and reflective of, their needs. By following such guidelines, people can learn to design, and handle with care, the incredible forms of power that lie in their hands—this aim is a crucial one if people desire to preserve the planet on which they live.

Keywords Cyberspace • Design • Fast • Guidelines • Limits • Planet • Slow • Slower • Speed • Time

In this chapter, we provide an introductory investigation to the limits of time and how ICT is fast changing human beings' concept of time as well as people's everyday lives in terms of their doing and their being. The chapter is about time and its relationship with technology and human beings. It is also about *limits*. It begins with an exploration of the myth

© The Author(s) 2018
N. Patrignani, D. Whitehouse, *Slow Tech and ICT*,
https://doi.org/10.1007/978-3-319-68944-9_3

of speed and its potential decline and new ways of examining the challenges of human limits.

In the physical world, human beings have limits: they need to respect the limits of a finite planet. This is the most important lesson that resulted from the ecological reflection process started by Rachel Carson in the early 01960s (Carson 1962) and by the report of Club of Rome in the 01970s (Meadows et al. 1972). This awareness emerged in a world still centred on thinking inherited from the Industrial Revolution, based on matter and energy.

However, what about the era human beings are entering now that is based on a Knowledge Revolution? Do people also have limits in the virtual world? Is there an ecology of cyberspace? What kind of new relationships—for example, between human beings and artificial agents—are people aiming to develop so as to ensure better forms of well-being and well-living?

Is ICT really improving people's quality of life? Can people try to identify some guiding principles or values that could influence the development of technologies from the point of view of society? Can one imagine an ICT that contributes positively to people's well-being and well-living without having negative side effects on human beings, on their social structures, and on the planet? This set of questions are ones that always need to be asked in any setting or situation in which ICT is to be used. They form a protocol for understanding Slow Tech.

By introducing the time dimension into technology, Slow Tech means that when people continue to use ICT in the future, they will do so in a more conscious, responsible, sustainable, and ethical way. People need to regain control of the pace of the days of their lives by designing technologies that are more respectful of their brain and body and are not necessarily based on a continuous increase in clock speed. Human beings should start looking at longer-term perspectives and at a new conception of well-being and well-living that is based on Slow Tech.

3.1   An Ecology for Cyberspace?

In order to find possible answers to such fundamental questions as whether there are any limits in the Information Society, or if there is an ecology of cyberspace, people need a more precise description of the new knowledge-based era. We describe it here as a Knowledge Revolution.

3.1.1   Characteristics of the Knowledge Revolution

A comparison can be drawn between the world view inherited from the Industrial Revolution and the new world view of the Knowledge Revolution. That initial world view is now decreasing in importance. The second is increasing in priority. During a transition phase, they will of course both continue to coexist. It is, nevertheless, possible to examine the eight main characteristics of the two world views side by side.

First, in the industrial period, the most important currencies were matter and energy. In today's knowledge-based world view, the most important currencies are information and knowledge.

Second, in the industrial age, conservation laws were based on the principles of physics, in which matter and energy were interchangeable and could only move from one form to another. It was impossible to destroy either matter or energy. However, there is no longer a conservation law. Conservation laws apply only to atoms, and not to bits of information. Now information can be both created and destroyed easily. When a support is a physical entity—for example, a stone, paper, a video, a compact disk, or a memory stick—capable of storing different forms of data, information is based on differences in that support, that is, it is matter and energy in an ordered state. As a result, power supplies are needed to maintain those differences (Patrignani et al. 2011). While conservation laws used to provide an assurance of stability, one can now lose easily the differences in the support. As a result, one can also lose the information itself. Information is more fragile and less secure. This is why people need to make backup copies and organise emergency supplies.

Third, in the past, the main systems that people designed were continuous linear systems. Now they design finite-state machines—systems or computers that can move between a discrete, enumerable number of states and, often, a gigantic number of states.

Fourth, designers used to be able to test linear systems through exhaustive stress tests. These tests solicited—that is, they put pressure on—a system up to an acceptable threshold. Beyond this threshold, the system collapsed. As a consequence, these thresholds provided the maximum levels of acceptable solicitation. Below such a level of solicitation, a system could be considered to be stable and reliable. Now designers cannot test their complex computer systems exhaustively—because there are no tolerances or thresholds. A change in a single bit could crash a large-scale system. Today designers have to limit their testing to the principal functions and accept instead the unavoidable unreliability of software. As Edsger W. Dijkstra, one of the founders of computer science, said: '... program testing can be used to show the presence of bugs, but never to show their absence!' (Dijkstra 1972: 7).

Fifth, in the former world of matter and energy, people's relationships were dominated by zero-sum exchanges (if I give you one Euro, you give me one Euro: therefore our wealth is not increased). Now people have positive-sum exchanges (if I give you one idea, you give me one idea, and our knowledge is increased).

Sixth, before the computing era, a copy was unavoidably different from the original. Now the copy is the original (a bit is a bit). Two exchanges, which took place over 180 years apart, illustrate the dynamics of this relationship between knowledge and information—they show that ideas and concepts often take a lengthy period to be absorbed into society at large and it takes time to innovate. Two centuries ago, on 13 August 01813, Thomas Jefferson wrote to Isaac McPherson about the intangible nature of knowledge: '... he who receives an idea from me, receives instructions himself without lessening mine; as he who lights his taper at mine, receives light without darkening me ... ideas should spread from one to another over the globe ... inventions then cannot, in nature, be a subject of property' (Jefferson 1813). More recently, the futures thinker, Peter Schwartz, wrote: 'Knowledge is the only kind of wealth that multiplies when you give it away' (Schwartz 1996).

Seventh, in a world that was dominated by matter and energy human beings experienced high environmental impacts as a result of their activities. In the new knowledge-based society, one can anticipate that this impact will be lighter.

Eighth, in the physical world, movements were restricted and speed is limited by friction. In the new world, human beings can move bits around at the speed of light: it is the only limit.

These eight concepts, and their contrasts with each other, are listed in tabular format in Fig. 3.1.

ICT represents a watershed; it has triggered a new information era. New tools are maybe also needed to understand the limits of ICT and their side effects on human beings.

After various different types of catastrophes, human beings are increasingly aware of what was right and what was wrong about the industrial era. However, how can people judge what is right and what is wrong with the new knowledge-based era? What are the connections between matter and energy, on the one side, and information, on the other?

Industrial Revolution		Knowledge Revolution
Matter and Energy		Information and Knowledge
Conservation Laws		No Conservation Laws
Continuous Linear Systems		Finite-State Machines
Exhaustive Stress Tests (tolerances)		Functional Testing (no tolerances)
Zero-Sum Exchanges		Positive-Sum Exchanges
Copy ≠ Original		Copy = Original
High Environmental Impacts		Low Environmental Impacts
Limited Speed (friction)		Speed of Light

Fig. 3.1 Characteristics of the industrial revolution and knowledge revolution

According to Norbert Wiener, the founder of cybernetics, information is a fundamental constituency of *being* in which people are simply whirlpools in a river of ever-flowing water—patterns that perpetuate themselves (Wiener 1954: 96).

A similar notion to Wiener's idea of being was used by the anthropologist and philosopher Gregory Bateson when he defined the concept of *mind* and its linkage with patterns (Bateson 1979: 8).

Both scientists focused on the fundamental links between human beings and the surrounding nature, its mutability and change and yet also its permanence, and the emerging importance of a new dimension: information. Today, there are similar trends in more recent and even contemporary theories of physics that propose a universe made up of information encoded in matter and energy (Wheeler 1990; Lloyd 2006; Bynum 2008).

These notions help to introduce the ideas of the infosphere. The infosphere was defined by Luciano Floridi as the whole information environment, constituted by all information entities (including artificial agents and offline spaces), their interactions and properties, and in which a physical object is also an information entity (Floridi 1999). From this concept arises an analogy between suffering in the biosphere and entropy in the infosphere (Floridi 1999).

What is it that people need to do in order to guarantee that there will be more focus on human beings and their quality of life when ICT is developed and used? Progressing from this understanding of human beings' role in their basic environment, and the synergy between beings and the surrounding climate, means there is a need to face fundamental ethical issues not only in terms of society at large but also of the information and the technologies they use.

3.1.2   Human Beings and the Concept of Time

Human beings have probably always been preoccupied by what precisely it means to be human. Certainly, the literature of Ancient Greece—with its reference to oral history—shows that this dilemma engrossed writers and thinkers since early times (Plato 2015). Much of this deliberation about humanity took place in its own right, but some considerations emerged in response to the introduction of technologies and the way in which this results in changes in society.

In terms of the twentieth century, concerns about what it means to be human were prevalent in the post-Second World War period and particularly the 01960s. Today, there is a revival and extension of many of the notions that emerged, or even re-emerged, around 50 years ago. Questioning who and what we are as human beings is once again on the rise (Floridi 2014; Dewandre 2014).

Time is very difficult to define precisely. Human beings use different scales and instruments according to their experiences. Even in theoretical physics, the question of what is time remains very open (Rovelli 2017). While time can be measured, from the scientific and philosophical points of view, its conceptual definition is one of the most complex challenges in contemporary research (Hawking and Mlodinow 2008). Despite this lack of an ontological definition of time, in every single recent year, human beings seem to feel that they are running faster and faster thanks to new ICT systems and production processes. Human beings are speeding up, trying to run increasingly fast not to go somewhere specific but just so as not to lose their current position.

Life is becoming a Formula One race, where the racing teams have to ensure that the cars minimise their maintenance time in pit stops simply in order to stay among the competition. As Geoff McGrath of McLaren has highlighted, tyres on racing cars can be removed and replaced in two seconds or less (FT Live 2015). In McGrath's opinion, 'going faster can help us live longer' (Rundle 2015). In contrast, as we have stated elsewhere (Patrignani and Whitehouse 2013), by taking time to reflect about ICT, Slow Tech can provide 'a route to general awareness and acceptance of the "brakes required in the technological Indianapolis"' (2013: 389).

Computers may be even changing human beings' notions of who they are and how they perceive themselves. Human beings risk experiencing a new kind of totalitarianism (Rosa 2010). At the core of this acceleration there are computer systems. ICT is the main engine for accelerating human lives and all the various associated production processes. Is it possible to develop a different kind of ICT? Is it possible to design ICT systems that instead put human beings at the centre?

It is, therefore, important to look both backwards in time and forward. This dual approach is well worth using at any important period or phase in time, to find a balance between the two sets of reflections. Janus, the god of beginnings and doorways/thresholds, is a symbol of the importance of shifting from one state to another.

Time is a convention that has been continually evolving throughout history. Human beings started to measure time in the very beginning of history when they looked at the sun rising every morning or at the stars or the moon at night. Indeed, people used to measure time in terms of moons (or new moons), a period of about 28 days. Yet, human beings modify the notion of time according to what is happening around them.

With the use of computers, people's concept of time is changing quite rapidly. People have started to reason in parallel. With computers human beings are multi-tasking: people think they can control several processes that are evolving in parallel. It seems to them that their time is 'compressed' and it can contain more than one task or scenario at any one moment. It looks as though computers increase people's productivity. Yet is there a limit to this growth? (Miller and Buschman 2015).

In the twenty-first century, human beings' notions of time are changing based on the growing influence of computers and physics. First, we will look at the common, day-to-day sense of how computers are modifying people's sense of time and how people have faced the challenge of changing time concepts throughout history. Second, we will progress to some more challenging examples that emerge from a set of scientific fields.

3.1.3   Why Human Beings Do Not Like Limits

As human beings, people do not like limits: they are constantly looking for new challenges. Historically, this is an old story. Myths and classical stories illustrate very effectively the ways in which the human race has faced such tests and trials. They offer indications of just how fundamental to human beings is the need to test themselves. Here are just two examples of mythical stories that have been taken from Ancient Greece: the stories of Prometheus and Ulysses.

One of the oldest myths in human history is that of Prometheus, who was one of the Titans in ancient Greek mythology. Prometheus was considered to be a real champion: he stole fire from the god, Zeus, to give it to mortals. The punishment enacted on Prometheus for this audacious crime was terrible. He was bound to a rock and, every single day, a great eagle ate his liver. Since Prometheus was particularly known for his intelligence, his daring act became the symbol of the search for new knowledge.

Another ancient story is that of Ulysses. Ulysses would not accept the borders of the known universe of his time (he challenged the geographic and navigational limits of the Pillars of Hercules, the promontories at the entrance of the Strait of Gibraltar). Facing up to the fears expressed by the crew on his boat, Ulysses urged the sailors to travel forward to make a courageous discovery. Ulysses' famous challenge is described in Dante's Inferno: 'Considerate la vostra semenza: fatti non foste a viver come bruti, ma per seguir virtute e conoscenza' (call to mind from whence ye sprang: ye were not form'd to live the life of brutes, but virtue to pursue and knowledge high) (Alighieri 1309, vv. 112–120). It is ongoing searches for new knowledge, like this, that boost the challenge to human limits.

Since people need challenges in their lives, they continue to make use of fascinating myths like these. Such ancient myths are still present in society and form the basis for the pursuit of scientific research.

Now, however, at the end of the Industrial Revolution and the dawn of the new Knowledge Revolution, human beings have in their hands the power to destroy planet Earth. People have gone well beyond their challenge to the limits of the Earth.

Now people have to become aware that many of the new challenges that they face lie physically inside themselves in embedded devices and gadgets. Computers are among the core engines that are speeding up many unsafe processes. People are living on a finite planet: their continuing exploitation of natural resources, based on processes that are increasing in speed, needs to be re-examined. Human beings have to incorporate the concept of limit, which is imposed by the current environmental and social crises, in all their plans and projects including new initiatives around ICT.

This is the new challenge in people's lives. Indeed, Grace Hopper, the female computer scientist who first proposed the use of high-level languages for computer programming, paraphrased many times a saying attributed to John A. Shedd in 01928: 'A ship in port is safe; but that is not what ships are for – sail out to see and do new things' (Hopper 1952).

Many of these new challenges that people are investigating are scientific in orientation. Examples—from physics, space, materials, and ICT—illustrate innovative ways of handling challenging aspects of speed, distance, and time.

First, speed: in physics, researchers are investigating elementary particles. Scientifically, according to the standard model theory in physics, there are some particles called quarks that exist below atoms, below the nucleus, and below protons. The decay of a quark is the spontaneous process of its transformation into other elementary particles. The decay of one of the known quarks, the top quark, requires 0.5 yocto-seconds that is 0.5×10^{-24} sec. Thus, there are phenomena in particle physics that—in order to be measured—need only a yocto-second (10^{-24} sec) order of magnitude.

In ICT, other artefacts increase the processing power of computers up to the speed of a single operation that takes less than 0.1 atto-seconds (10^{-18} sec): in the exponential supercomputer competition, Fuijtsu's K-Computer runs at more than 10,000 TeraFLOPS (floating point operations per second), that is, more than $10,000 \times 10^{15}$ FLOPS, or 10×10^{18} FLOPS, which means that one operation needs less than 0.1×10^{-18} sec or 0.1 atto-seconds to be executed (Nakamura 2011). Until 02015, the fastest computer on the planet was Chinese and was called Tianhe2; it ran at more than 33,000 TeraFLOPS (TOP500 2013) at China's National Computer Centre, Guangzhou, China.

Second, in terms of distance and size, through space exploration, astronauts and machines have started transporting artefacts afar. They are being sent from the Earth to the moon, to Mars, to deep space, to the distance of yotta-metres (10^{24} m) of the known edge of the universe. The diameter or dimension of the observable universe is estimated by current cosmologists to be around 900 yotta-metres, that is, 90 billion (90,000 million) light years, where one light year is about 10^{16} metres.

In new materials investigation with nanotechnologies—which head in completely the other direction towards the minute or small—scientists and developers are manipulating matter at the scale of nanometres (10^{-9} m).

Third, in the time dimension, people are now building artefacts that are planned to last for thousands of centuries, many Tera-seconds (10^{12} sec). The Onkalo nuclear waste site in Finland, for example, is planned to last for 100,000 years, which is about 3.1 Tera-sec (10^{12} sec). We explore more examples along these lines in Chap. 4 of this book.

If human beings need to overcome both their own limits and those of the planet, Slow Tech provides a fascinating challenge for people in the ICT world. Slow Tech is in reality an initiative to search for new knowledge to create, design, and make a wise use of ICT.

3.2   THE MYTH OF SPEED

Speed in itself is just a physical variable. Sometimes it is useful to act fast, for example, when health workers must bring the victim of an accident or a patient to the hospital in an ambulance, save a child from falling from a chair, or simulate via a computer the behaviour of an incoming hurricane and prepare for an evacuation. When people, and even society as a whole, are at risk, it is always important to act quickly. In this way, one can help to ensure the protection of societal infrastructure and frameworks.

In the twentieth century, however, speed became a myth. Every single human activity began to look better if it was faster.

At the beginning of the last century, the coupling of modernity and speed lay at the core of the Futurist movement. At its launch, Futurism was considered to be at the avant-garde of culture. Indeed, the Futurists wrote that: '...the splendor of the world has been enriched by a new beauty: the beauty of speed' (Marinetti 1909). For the Futurists, progress and modernity were synonymous with speed. Unfortunately, this beauty of speed was based implicitly on the conviction that human power would overtake and dominate nature.

This desire for domination formed the cultural background to the dictatorships of the last century. People now know that this illusion of the control of nature was a terrible mistake. Going faster can offer the illusion of controlling the surrounding reality. ICT takes this illusion to the extreme: with just a click people can quickly change the (virtual) world in front of their eyes.

Today, however, in contrast, human beings are aware that the complexity of natural systems is so intense and delicate that it is better to find new ways to coexist with these systems instead of dominating them.

Speed is no longer a value in itself. Instead, people are discovering that, before initiating new projects and initiatives, they need to take more time. They are aware of the need to consider the impact that new advances might have on natural and social systems, and maybe they even need to locate alternative solutions or different projects. Hence, they need to slow down (Giuffreda 2011).

In order to cope with the challenges that face them, people need innovations. Yet, they must also innovate in a way that enables them to use the incredible power in their hands with care and sensitivity, without destroying the living environment.

ICT is at the core of the increasing speed of society's processes today. Yet does one always really need faster and faster computers? Or is it time to rethink computer speed? Should one introduce computers in society, not just because they are available but when and where they really make sense? Should one not make wiser uses of computers by allocating enough time for human relationships and for the reflections needed by human minds?

REFERENCES

Alighieri, D. (1309). *Inferno*, CANTO XXVI, Vol. I (A. M. Rev. Henry Francis Cary, Trans.). London: Printed for Taylor and Hessey, Fleet Street. 1819.

Bateson, G. (1979). *Mind and Nature*. New York: Dutton.

Bynum, T. W. (2008, September 24–26). The Nature of Universe, Human Nature and Contemporary Information Ethics. In T. W. Bynum, M. Calzarossa, I. De Lotto & S. Rogerson (Eds.), *Living, Working and Learning Beyond Technology*, Proceedings of ETHICOMP 2008, University of Pavia Press, Mantua.

Carson, R. (1962). *Silent Spring*. Boston: Houghton Mifflin.

Dewandre, N. (2014). *Policy-Making in a Hyperconnected Era: Game Over for Modernity?* European Commission. http://ec.europa.eu/digita-agenda/en/onlife-initiative. Accessed [9 October 2017].

Dijkstra, E. W. (1972). *Notes on Structured Programming*, 1972, Section 3, On the Reliability of Mechanisms. EWD249.

Floridi, L. (1999). Information Ethics: On the Philosophical Foundation of Computer Ethics. *Ethics and Information Technology, 1*, 37–56, Kluwer Academic Publisher.

Floridi, L. (2014). *The Fourth Revolution: How the Infosphere Is Reshaping Human Reality*. Oxford: Oxford University Press.

FT Live. (2015, June 10). *FT Digital Health Summit Europe*. https://live.ft.com/Events/2015/FT-Digital-Health-Summit-Europe. Accessed [9 October 2017].

Giuffreda, G. (2011, May 27). Elogio della lentezza. *Il manifesto*, Roma.

Hawking, S., & Mlodinow, L. (2008). *A Briefer History of Time*. London: Bantam Press.

Hopper, G. (1952). *The Yale Book of Quotations* by Fred R. Shapiro, Section: John A. Shedd, Page 705, New Haven: Yale University Press, 2006. John Augustus Shedd, *Salt from My Attic*, Portland: The Mosher Press, 1928 quoted in 'Grace Hopper: The Youthful Teacher of Us All', by Henry S. Tropp in *Abacus, 2*(1) (Fall 1984): 7–18.

Jefferson, T. (1813). *The Writings of Thomas Jefferson* (Vol. 1–19, A. E. Bergh, Ed.). Washington, DC: Thomas Jefferson Memorial Association, 1905.

Lloyd, S. (2006). *Programming the Universe: A Quantum Computer Scientist Takes on the Cosmos*. London: Jonathan Cape.

Marinetti, F. (1909, February 20). Le Futurisme. *Le Figaro*. Published also by several Italian newspapers as *Manifesto del Futurismo*.

Meadows, D. H., Meadows, D. L., Randers, J., & Behrens, W. W., III. (1972). *The Limits to Growth*. New York: Universe Books.

Miller, E. K., & Buschman, T. J. (2015). Working Memory Capacity: Limits on the Bandwidth of Cognition. *Daedalus, 144*(1), 112–122.

Nakamura, A. (2011, June 20). Japan's K Computer Becomes Fastest Supercomputer in the World. *Technology Headlines*.

Patrignani, N., & Whitehouse, D. (2013). Slow Tech: Towards Good, Clean, and Fair ICT. In T. W. Bynum, W. Fleischman, A. Gerdes, G. M. Nielsen & S. Rogerson (Eds.), *The Possibilities of Ethical ICT*, Proceedings of ETHICOMP 2013—International Conference on the Social and Ethical Impacts of Information and Communication Technology (pp. 384–390), Print & Sign University of Southern Denmark, Kolding.

Patrignani, N., Laaksoharju, M., & Kavathatzopoulos, I. (2011). Challenging the Pursuit of Moore's Law: ICT Sustainability in the Cloud Computing Era. In D. Whitehouse, L. Hilty, N. Patrignani & M. Van Lieshout (Eds.), *Social Accountability and Sustainability in the Information Society: Perspectives on Long-Term Responsibility*, special issue of *Notizie di Politeia, 27*(104).

Plato. (2015). *Apology*. Translated by Benjamin Jowett. MIT. Internet Classics Archive. http://classics.mit.edu/Plato/apology.html. Accessed [9 October 2017].

Rosa, H. (2010). *Alienation and Acceleration: Towards a Critical Theory of Late-Modern Temporality*. Aarhus: Aarhus University Press.

Rovelli, C. (2017). *L'ordine del tempo*. Milano: Adelphi.

Rundle, M. (2015, April 24). McLaren's Geoff McGrath: Going Faster Can Help Us Live Longer. *Wired Events*. http://www.wired.co.uk/news/archive/2015-04/24/geoff-mcgrath-mclaren-applied-technologies-wired-health-2015. Accessed [9 October 2017].

Schwartz, P. (1996). *The Art of the Long View*. New York: Currency Doubleday.

von Schomberg, R. (Ed.). (2011). *Towards Responsible Research and Innovation in the Information and Communication Technologies and Security Technologies Fields*. European Commission, Directorate General for Research and Innovation, Luxembourg, Publications Office of the European Union. http://ec.europa.eu/research/science-society/document_library/pdf_06/mep-rapport-2011_en.pdf. Accessed [9 October 2017].

Wheeler, J. A. (1990). Information, Physics, Quantum: The Search for Links. In S.-'i. Kobayashi et al. (Eds.), *Proceedings of the 3rd International Symposium Foundations of Quantum Mechanics in the Light of New Technology, 28–31 August 1989* (pp. 354–368). Tokyo: Physical Society of Japan.

Wiener, N. (1954). *The Human Use of Human Beings: Cybernetics and Society*. Boston: Houghton Mifflin, 1950; 2nd edn revised, Garden City: Doubleday Anchor, 1954; citations are from the 2nd edn revised, 1954.

CHAPTER 4

Stories About Speed and Time

Abstract In this chapter, the focus is on perspectives about the time needed for change. Five case studies concentrate on human limits, historical developments, speed, and time. For example, the Lotka-Volterra model which examines the relationship between prey and predators indicates just how long-term phases and developments can be. A longer-term view highlights the need to protect human existence: by burying nuclear waste in Finland or safeguarding seeds and crops for millennia in Norway. The Long Now Foundation proposes that adding a zero in front of the year in which humankind is living can make time look quite different—a formula that, with the exception of the dates of publications, the authors adopt throughout the whole of this book. Looking at time through such a longer-term perspective can encourage people to zoom in and focus on items and issues that are really important within the longer timeframe of survival and sustainability. What happened on the New York Stock Exchange in May 02010 indicates just how out-of-human control technology has already become, particularly in the financial field. As a result of using a case study-based approach that emphasises a longer, slower understanding of time, as in this chapter, awareness can grow of how much more needs to be done to both gain and regain control over critically important information systems.

Keywords Case studies • Change • Control • Cycles • Future generations • Long Now Foundation • Lotka-Volterra model • Long-term • New York Stock Exchange • Nuclear waste • Planet • Time

To introduce a set of arguments about why speed is not always important or necessary and the concept of time is a pure convention, five case studies have been selected. Each comes from a different location around the globe. Four of them are from real life.

The first case study is simply a model. The Lotka-Volterra model shows the balance that occurs between preys and their predators. It provides a fundamental reminder of how different development can look when it is examined over a sufficiently long period of time. People are beginning to think about very long time periods and, indeed, phases that may extend beyond the existence of human beings on this globe.

The following three case studies draw long-term planning to our attention: they each raise the challenge of how to alert future generations about the location of the materials and equipment that they contain, whether for good or bad reasons. The second case study comes from Onkalo in Finland: it holds long-term danger since it is a storage space for nuclear waste. The third case study, based in Svalbard, off the Norwegian coast, houses beneficence because it stores examples of plant seeds for long-term future use. The fourth case study, in Van Horn in Texas, is also a location with a benevolent message: it describes a clock which is intended to run for at least 10,000 years, buried deep in the desert.

The fifth case study alerts people to the dangers inherent in a reliance on computers to run critical systems pushed to their extreme speed: it is the example of the 02010 flash crash when it became frighteningly apparent that human beings were no longer in control of the financial systems on the New York Stock Exchange.

4.1   THE LENGTH OF LIFE CYCLES: THE LOTKA-VOLTERRA MODEL

Human beings do not like limits, but they do have to accept that they are small animals on a single, limited planet. When people look around themselves and try to analyse a number of physical variables with a view to their evolution over time, they find that all functions are limited. Maybe they observe some apparent form of exponential growth. However, sooner or later, all growth will be saturated and reach its limit. It looks as though, on planet Earth at least, only stationary systems can survive in the long run.

The well-known Lotka-Volterra mathematical model describes a simple system formed by two species: one is the prey (e.g. small fish) and the other is the predator (e.g. large sharks) (see Fig. 4.1). Lotka-Volterra's prey-predator model, described by the American demographer Alfred J. Lotka in 01924 and by the Italian mathematician Vito Volterra in 01926, is a linear system of differential equations of the first order.

One can make the following simple assumptions from this model: the population of preys increases exponentially in the absence of predators, and the population of predators decreases exponentially in the absence of prey. If the predators consume too many prey, they will have less food, and so the population of predators will decrease. The population of prey can then increase again, and finally the population of predators will have

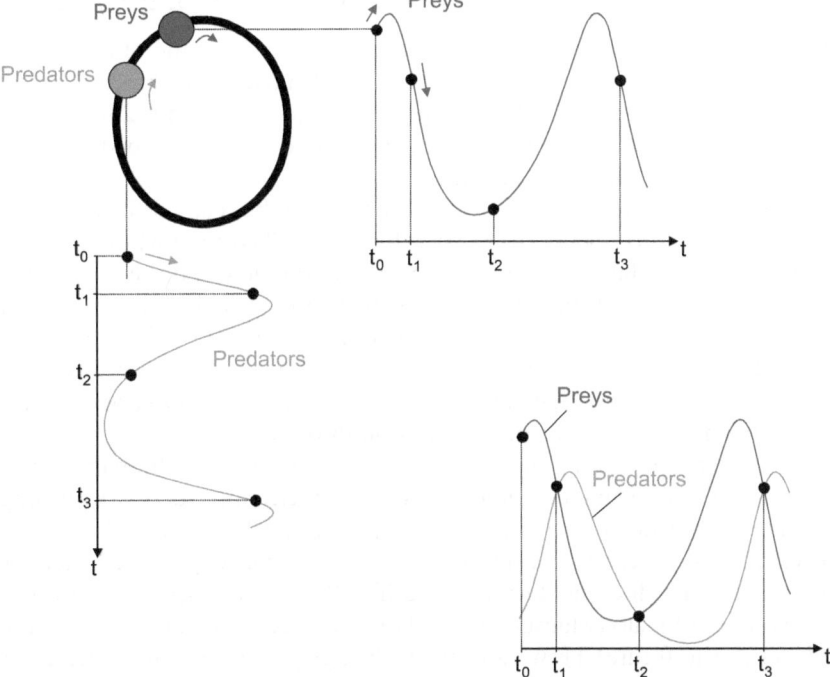

Fig. 4.1   Lotka-Volterra's prey-predator model

more food. It is a cycle. Of course, however, sometimes something goes catastrophically wrong, and extinctions occur.

On planet Earth, nature is composed of a collection of cycles (e.g. the carbon cycle, water cycle, and life cycle). In nature only processes that can be represented by cycles can survive in the long term. Life itself has probably been possible on this planet since natural evolution developed cycles at a molecular level, such as the fundamental cycle that generates chemical energy from carbohydrates, fats, and proteins: the Krebs' Cycle inside mitochondrial cells (the 'powerhouse of the cell') (Krebs and Weitzman 1987).

Why are such cycles important for reflections about time and the long-term view? They are important because they show human beings that exponential growth is unsustainable. Up until today, ICT represented the best example of exponential growth (in speed, power, etc.), but the deep question arises: is ICT sustainable in the long term?

4.2   Dangers Are Long Lasting: Onkalo, Finland

Onkalo in Finnish means a hiding place. It is also the name of the facility that will act as the long-term storage for all nuclear waste produced in Finland, a few miles from the Olkiluoto nuclear plant, about 280 kilometres north-west of Helsinki. This cave, encased by granite bedrock and constructed in a spiral shape, will be 520 metres deep.

The construction started in 02004, and it will be ready to host nuclear waste by 02020 (Onkalo 2015). It will then be filled for a whole century, until 02120. At that point its mission will be complete and the tunnel that leads to the cave will be refilled and sealed forever. This is a project that was started while a generation of people from the twenty-first century is alive, but its job will be complete when all those living today will have been dead for a very long time. The nuclear waste that Onkalo contains will remain dangerous for more than 100,000 years.

Onkalo is opening up new kinds of questions related to human beings' concept of time. Some technologies, such as nuclear products, are pushing people in directions for which they simply do not have the philosophical background to deal with a number of profound challenges. Into Eternity is the title of a documentary released in 02010 that poses difficult questions about Onkalo (Madsen 2010). For example, what will become of the planet in the future? How can human beings prevent future generations from entering into direct contact with Onkalo's contents? What kinds of messages, languages, or symbols can be used to communicate Onkalo's underground danger?

Human civilisation has existed for just a few thousand years (e.g. the Egyptian pyramids are just 05000 years old). People's minds cannot easily make projections about long-term futures. The human concept of time is deeply affected by the artefacts built by human beings and the technological choices that they make. The relationship between human beings and time is one of the oldest ethical issues of humankind and has fascinated people from their very origins.

Onkalo is an excellent example of an attempt to predetermine the future. Yet a hundred millennia on the human scale is close to eternity, the *saecula saeculorum*. Thus, imagining the implications raised by this underground storage space extends beyond scientific norms. The developments taking place at Onkalo are precisely where the zeal to dominate nature crosses the boundary of scientific acceptability and enters the domain of magic. It would probably be wiser to accept the experience of limits that is strongly associated with the human body and human physical and intellectual experience. Yet, Onkalo is at least one of the few initiatives that is addressing in an acceptable way the problem of nuclear waste management.

In Europe, in Germany and Italy, there have been political decisions made over the past several years that are related to the closure of various nuclear power plants and to stopping the production of nuclear waste. They indicate that the dangers posed by a number of nuclear disasters (e.g. Three Mile Island, United States of America 01979; Chernobyl, Soviet Union 01986; Fukushima, Japan 02011) have in certain cases been transformed into learning experiences.

For some politicians at least, their vision of the future is now based more on a concern and consideration for saving and reducing energy consumption and developing renewable and cleaner sources (e.g. hydroelectricity, geothermal power, wind, solar power, biomass, and the oceans' waves). Strikingly, all these new cleantech solutions are based on renewable, sustainable, and long-term cycles.

4.3   ETERNALLY SAVING CROPS FOR HUMANITY: SVALBARD, NORWAY

Svalbard is the name of a Norwegian archipelago about 3000 kilometres north of Oslo. William Barents discovered it in 01596. In the seventeenth century, it was used by whaling and fishing boats but was afterwards abandoned. Its underground resources led to coal mining at the beginning of the twentieth century. In 01925, these remote islands became part of Norway.

The islands have very few roads, and the people who live there or visit them move around by using snowmobiles or ships. Travelling around one of the islands, one eventually comes to a strange construction with a small door: this is the Svalbard Global Seed Vault, the seed bank for eternity. The entrance to this vault provides access to a 120-metre tunnel under a sandstone mountain that is equipped for cooling seeds to −18°C. It is a place that is able to preserve about 4.5 million seeds for hundreds of years to come so that some grains will be able to survive for even longer, perhaps for thousands of years.

The construction was built with the mission of saving the most important seeds on planet Earth. The goal of the seed bank is to maintain some reserves of seeds in the event that, for some reason, all other seed samples were destroyed. Indeed, the Svalbard facility stores duplicates of seeds from several other collections around the world (many of them are in developing countries). If these collections were to be lost, they could later be regenerated with the seeds from the seed bank (MAAF 2011).

The idea of protecting seeds from possible catastrophes is not new. People have now started to design artefacts that could survive doomsday-like events. These events used to be the consequences of natural events such as volcanoes, asteroids, or earthquakes. Yet, today there are additional artificial catastrophes such as nuclear threats, bioterrorism, and global warming. Since the seed bank is 130 metres above sea level, it could also survive the rising sea level caused by global climate change.

The store, which was begun in 02006, has cost about $9 million. The Norwegian government initiated its building. For this purpose, it is supported in terms of operational costs by a combination of international organisations and foundations. The seed bank was officially opened in 02008, and its management is ensured by a three-party alliance: the Norwegian government, the Global Crop Diversity Trust, and the Nordic Genetic Resource Center.

The idea of this seed bank is the very opposite of a nuclear waste management site, a location from which one would want future generations to be kept away at all costs. With such a seed bank, one would want future generations (or at least the human beings who survive a doomsday event) to be able to enter the place and gain access to the seeds so as to be able to sow them in the ground.

Once again, human beings are challenging the limits of time. Curiously, the official Svalbard website announces that '... The facility is designed to have an almost "endless" lifetime'. Svalbard provides people with yet another time game.

As with the Onkalo case, in Svalbard, human beings will need to ensure information exchange with future generations. Will ICT be able to help in this very long-term information exchange?

4.4   STIMULATING LONG-TERM THINKING: VAN HORN, TEXAS

Hiking for an entire day over the mountains of western Texas close to Van Horn, one eventually encounters a construction site. Inside this mountain in the United States of America, a group of visionary people who are united in the Long Now Foundation is building the first monumental clock, 'The 10,000 Years Clock'. The clock itself is huge: it is more than 50 metres high.

Once finished, the clock will be open to visitors who will have to climb a spiral staircase for hundreds of metres in order to see the huge mechanism that drives the clock ... it is the slowest computer in the world. This complex system will capture and store mechanical energy as a result of the physical moves that visitors make when inside the building.

However, the clock is also designed to run on its own for 10,000 years. It stores mechanical energy by using a smart system that accumulates different temperatures captured from the rays of the sun (there is a hole at the end of a vertical tunnel at the top of the clock). This thermal power is translated into mechanical power by transmitting heat inside the clock by means of long metal bars. So, even if no one comes to visit the clock, the sun will power it.

The clock is also able to make self-adjustments: it will use the power of the sun to correct small variations in the length of the day. At noon the sun will lie exactly over the tunnel, and a prism will direct its rays down inside to be captured by the mechanical engine and interpreted as a synchronising signal.

The engineering challenges in building the clock are considerable. The clock will, for example, play different melodies and is programmed not to repeat any music for 10,000 years.

Yet, the clock's real goal is to stimulate long-term thinking. It will help people's thinking develop over a long period of time. The main question is to people as human beings: how are people acting now vis-à-vis future generations? Are we being good ancestors? (TLNF 2015).

The Long Now Foundation is at the basis of the development of this clock and other projects. It has as its mission the goal of enlarging the concept of time: 'to provide a counterpoint to today's accelerating culture and help make long-term thinking more common' (TLNF 2015).

The fascinating fact is that most of the people involved in this initiative are people who come from high-tech domains. They include Danny Hillis (a parallel computing pioneer, the designer of the clock); Stewart Brand (founder of the Global Business Network, *Whole Earth Catalog*, the Hackers Conference and The WELL); Esther Dyson (pioneer of the digital age and founder of ICANN); Kevin Kelly (founder of *Wired* magazine); and Mitchell Kapor (pioneer of personal computing and who established the software company Lotus). It looks as though this group of people is expressing a need: the need to zoom out, from the narrow time of computer clocks that run at the scale of nanoseconds, to the immensity of a 10,000-year project. It seems like they want to show a more responsible face to future generations by building the slowest computer on planet Earth.

The Long Now Foundation also has an interesting approach to the use of dates. The use of two digits instead of four digits to represent the years was a common programming practice in the mid-twentieth century, due to scarcity of computer memory. The problem became serious as the twenty-first century approached. The year 1999 (represented by '99') could roll over to '00', that is, back to (0)1900 rather than forward to (0)2000, which is why it was called the millennium bug. When writing dates, therefore, the Foundation uses five digits for each year, for example, to identify 2017, 02017 is written. This is also the convention adopted throughout this book.

Up until now, ICT has pushed human beings towards a very short-term view of time. This simple five-digit suggestion to enlarge the historical view at least up to 10,000 years is helpful in modifying people's behaviour towards lengthy goals and views of time. Such an approach helps symbolically and practically to emphasise the need to consider time in the long term.

4.5   Limiting the Speed of Machines: The New York Stock Exchange

Some seven years ago, on Thursday, 6 May 02010, something strange happened on the frantic floor of the New York Stock Exchange. The stock market was down just 1.5 per cent from the previous day. At around 2:30pm there was an exceptional event. At 2:47pm the rate was −9.16 per cent. In just ten minutes, a billion dollars had been lost. In only a few minutes, the indices that enumerate the market status had dropped hundreds of points. Some shares lost 90 per cent of their value (the value of shares in Accenture, e.g. fell to one cent), others were traded at incredibly high prices (Apple shares were purchased at $100,000). Operations were cancelled and the entire financial system stopped. What happened was later called the flash crash.

For the first time ever, human beings were not in control of the stock markets; computers controlled them. Indeed, the trading speed reached that May afternoon was simply impossible for human beings to follow, whether they were investors or brokers: yet everything was fine for the fast-trading software algorithms that were competing with each other!

It took months for the United States' Securities and Exchange Commission to produce a detailed report explaining the crash. Yet one thing was clear. The cause of the catastrophe was not human beings. It was machines, algorithms, and sophisticated software programs that were able to place 10,000 bids per second—through computer trading. These algorithms are unavoidably very similar. So, one algorithm emulates the behaviour of others that, in turn, influence others. This is the most effective, and dramatic, way to trigger a snowball effect. Machines can keep millions of variables under their control simultaneously. They are able to trade thousands of stocks at the same time in order to exploit infinitely small changes in stock prices. Every day more than 70 per cent of the volume of transactions are now controlled by computers that have taken over the trading industry (Salmon and Stokes 2011). At these speeds, cause-and-effect explanations of market phenomena become impossible.

Mary Schapiro, the chairwoman of the Securities and Exchange Commission, admitted that humans may need to reclaim some form of control from computers '… automated trading systems will follow their coded logic regardless of outcome' (Salmon and Stokes 2011). As a result, the Commission imposed circuit breaker mechanisms that slow down the speed of transactions (e.g. by automatically halting trading if the fluctuations are higher than ten per cent over a period of five minutes). All these measures go in the direction of slowing down algorithms.

So, it has now been acknowledged that the most sophisticated stock market in the world, controlled by the most advanced and fastest computer systems on the planet, needs to be reduced in speed. This is the admission that human beings now have to impose some limits on the speed of machines. People are currently conscious that they have delegated to incredibly fast operations and procedures—impossible for a human being to control—and stock market mechanisms that can change the destiny of people, entire companies, markets, and even society as a whole.

4.6   THE RECOVERY OF CONTROL

Human beings have created developments for which they now need to recapture control. Does this situation sound familiar? Particularly the fifth case study in this chapter may remind people of other, literary, examples. Two come immediately to mind.

In Goethe's 01798 poem, *Der Zauberlehrling* (The Sorcerer's Apprentice), exploiting an early form of automation, the young apprentice of an elderly sorcerer tries to use magic to enchant a broom that will fetch him water. Very soon, however, the apprentice loses control of the broom: unfortunately, he does not know how to stop it (Goethe 1798). Or perhaps people may think of Mary Shelley's 01818 novel *Frankenstein*. In this story, the unnamed creature created by Dr Victor Frankenstein causes a series of tragedies. Only at the very end of the story is the scientist who created the artificial life form horrified by what he has done (Shelley 1818).

All these case studies and stories show that it is important to keep in human hands the responsibility for processes and projects that can have long-term consequences for society and the planet. Preoccupation with the short term permits people to forget that they are integrated into much larger planetary cycles related to life, water, and carbon. Two simple illustrations can remind people instead of the importance of the longer term.

In Cambridge, England, the roof struts of the fourteenth-century colleges are built of oak. Acorns were planted in the college grounds to grow into oak trees. Thus, the founders of the university colleges were already thinking some 300 years ahead: they wanted the people of the future to be prepared for the time when the roofs would need to be repaired (Brand 1994). Similarly, it is said that—traditionally—native American Indians, when making major decisions, would determine what the meaning of any dilemma would be for their descendants up to seven generations into the future: 'In every deliberation, we must consider the impact on the seventh generation ... even if it requires having skin as thick as the bark of a pine' (Lyons 1980).

Overall these five case studies and ensuing stories can be seen as a set of impressive examples of the strong connections between long-term thinking and benevolent outcomes: when human beings start looking for a future for the planet and for future generations, they also start thinking in long-term cycles. Slow Tech is our way of introducing this long-term thinking into ICT.

REFERENCES

Brand, S. (1994). *How Buildings Learn: What Happens After They're Built*. New York: Viking Press.

Goethe, J. W. (1798). *Der Zauberlehrling*. https://de.wikisource.org/wiki/Der_Zauberlehrling. Accessed [9 October 2017].

Krebs, H. A., & Weitzman, P. D. J. (1987). *Krebs' Citric Acid Cycle: Half a Century and Still Turning* (p. 25). London: Biochemical Society.

Lyons, O. (1980). From the Constitution of the Iroquois Nations, The Great Binding Law, in 'An Iroquois Perspective'. In C. Vecsey & R. W. Venables (Eds.), *American Indian Environments: Ecological Issues in Native American History* (pp. 171–174). Syracuse: Syracuse University Press.

MAAF. (2011). *Ministry of Agriculture and Food, Svalbard Global Seed Vault Official Website*. https://www.regjeringen.no/en/topics/food-fisheries-and-agriculture/landbruk/svalbard-global-seed-vault/id462220/. Accessed [9 October 2017].

Madsen, M. (2010). *Into Eternity: A Film for the Future*. https://www.imdb.com/title/tt1194612/?ref_=nm_ov_bio_lk1. Accessed [9 October 2017].

Onkalo. (2015). *Official Onkalo Website*. www.posiva.fi. Accessed [9 October 2017].

Salmon, F., & Stokes, J. (2011, January). Algorithms Take Control of Wall Street. *Wired*.

Shelley, M. W. (1818). *Frankenstein, or the Modern Prometheus* (1st ed.). London: Lackington, Hughes, Harding, Mavor & Jones.

TLNF. (2015). *The Long Now Foundation*. http://longnow.org. Accessed [9 October 2017].

CHAPTER 5

Information: Environmental and Human Limits

Abstract This chapter investigates how information sharing and transmission have modified substantially notions of both space and time. Today, people are experiencing a sense of rapid expansion in information availability and use: their capacity to cope with this growth and the processing of data is often challenging for them. The urgency, importance, and time over which specific information is needed can modify how data are stored. Computers speed up the transfer of data, and how to store data in the long term is a dilemma. Just like crops and seeds, the long-term preservation of literary, artistic, and musical masterpieces is also important. It is now vital to focus on the ways in which processing power could be used to improve people's quality of life, well-being, and sustainability. This fifth chapter therefore explores a number of potential challenges that may limit human beings' activities: data transfer and conservation, human sensory and intellectual bandwidth, and sustainability.

Keywords Bandwidth • Characteristics • Chips • Collective • Computing • Dimensions • Entropy • Environment • Human • Individual • Limits • Moore's law • Provision • Sustainability

© The Author(s) 2018
N. Patrignani, D. Whitehouse, *Slow Tech and ICT*,
https://doi.org/10.1007/978-3-319-68944-9_5

Throughout human development, some activities have changed vastly while other aspects of people's lives remain substantially the same.

Information sharing in space and time is no longer the same. How human beings have transmitted information has altered considerably over the time period from 03500 BC until the present day, even if there is a return contemporarily towards a more aural tradition. Today, people are faced by a perceived explosion in information availability and its use. How urgent or important the information is, and the length of time over which it is needed, is changing the way in which that information is stored.

While computers now accelerate the transfer of data, there are challenges to data storage in the long term. The challenges are considerable. Just one ongoing dilemma lies in the preservation of literary, artistic, and musical masterpieces that people would like to conserve for future generations.

Another important imperative is the focus on the ways in which computing processing power could be used to improve the quality of life, well-being, and sustainability. Can contemporary progress in processing power be sustained? Can the computers that facilitate this data processing continue to be produced in ways that are sustainable for the planet?

Reflecting on these kinds of preoccupations raises people's awareness of how computers affect human behaviour. These considerations draw attention to the many more changes on the horizon that are related to data: for example, coexistence with artificial agents and with nano-bots and implantation of different forms of technologies inside human bodies. These developments too will have influences on what human beings do. Ostensibly, they may also modify who human beings are and what they desire to be.

This chapter introduces a discussion about the acceleration that ICT development has caused not only to the production of digital information but also to its sometimes indiscriminate dissemination (i.e. communication). These trends often contrast with human psychological and neurological constraints on filtering these stimuli and attending to them.

5.1   THE CHARACTERISTICS OF INFORMATION

It is crucial to delve deeper into an investigation of the term information. Information is used as a foundation stone or founding concept of contemporary society—the Information Society—but, as a term, can it be defined more precisely? The discussion loop that follows explores the question of what information is.

There are five main characteristics of information today. First of all, conservation laws no longer exist since information can be destroyed, created, improved, or refined. Second, an absence of information is a similar form of information to its presence: this is the famous nil of programming languages. Third, everything can be represented in bits: it is sufficient to establish an appropriate code to be able to do this, since code is a purely arbitrary convention. Of course, information provides the intangible side of reality or, more effectively, the representation of reality. While it is important to realise that people build models of reality on computers, these are, however, just models. To recall the pronouncement of the philosopher and scientist Alfred Korzybski (1879–1950): 'The map is not the territory'. Fourth, information does not exist without a form of support, yet information cannot be reduced to support: 'a hard disk is a hard disk is a hard disk', which is a paraphrasing of Gertrude Stein's 01913 statement that a 'rose is a rose is a rose'. Fifth, once information is based on a support of some form, it can be stored, processed, transmitted, and received by computers (Longo 1991).

Information is not an entity, it is a relationship. A humble description of this statement can be attempted: information does not exist on its own, and its support alone is not enough. Rather another entity is needed, an 'observer'. If, and only if, an observer or receiver is capable of detecting differences on the support (Bateson 1972) can the triggering of a complex interaction be detected: this occurs between the observer, that is, among her senses, her brain and mind, and the differences in the support. This complex interaction between the support and the observer can form the basic definition of information.

Once people store bits of information, can they assume that the bits will last forever? There is a risk that human beings may experience revenge on the part of the physical world! Bits are necessarily stored on a support, but any medium is fundamentally degradable. Even if one expects that the support will last forever, can it be supposed that any available input or output devices, that is, peripheral units, will be able to read the information? Even if one takes it as given that the information can be read, can it be guaranteed that the formats of the data can be recognised or understood: how are the bits packaged on the support and what code is used to write them? Even if one takes for granted that someone will know the formats, what if the data that are really interesting are encrypted?

Computers accelerate the rate of information exchange, yet there is also a high risk of the reverse effect: there is a danger that people will not be able to read computer-stored records in the long term. As a result, this could impede a cross-generational flow of information. Very soon a new set of skills will be needed by computer professionals who will need to be data archaeologists.

Regarding the content of the information, it is equally important to provide a context for the message or information that allows a receiver to infer as unambiguously as possible the sender's intended meaning.

These processes need time. Human beings need more time to reflect on information, particularly when the receiver needs to work at constructing or reconstructing the circumstances—era, culture, other contextual factors—that existed when the information was first sent or recorded. This challenge provides yet another incentive for taking a Slow Tech approach.

5.1.1 An Initial Focus on the Time Dimension

Computers enable human beings to go beyond the limits of space in information sharing, but what about time?

One of the major challenges that both single users and large organisations' infomasters (the persons in charge of the data management in an organisation with a large information technology infrastructure) are facing today is the information explosion. When planning the evolution of technological environments, one should carefully analyse human needs in terms of tools and storage requirements. Nowadays, with cloud computing, storage capability can be increased on demand.

One useful approach to information management, in particular when dealing with forms of information that are market based, is to bear in mind the parameter of an information lifetime. The information lifetime can be defined as the time it takes for information to lose its value. This can be caused by the delivery mechanism, through consumption or as a result of saturation.

For example, a typical daily newspaper has a lifetime of one day, due to its delivery mechanism. However, for an historian, a newspaper is a much richer artefact and will maybe become a useful reference. With a movie, the volume of all possible customers in an audience will typically be saturated after one or two years. The film will then be offered as a video or for online streaming. The market for this information disappears until a new set of customers is created as the result of an anniversary of the event on which the film is based or the subject matter handled in the film.

The choice of information container can be made on the basis of the information lifetime. For example, are users making very important decisions that must be taken in seconds? Or do they simply need to brainstorm? There are some situations where people must share information in a short period of time, such as in just few seconds; in this case, push technologies are useful for distributing very important announcements or news. However, in many other cases, real-time collaboration tools—such as audioconferencing or videoconferencing and application sharing—can satisfy requirements that are needed in the same time/different place (see, e.g. in the top left-hand box of Fig. 5.1).

An email message has a typical lifetime of several days; that is, it is assumed that if someone sends a message to a colleague, it will be read in a maximum of one or two days. In some organisations emails are followed up, and, if not replied to immediately, the workers are then sent a short message or receive a telephone call. The messages contained in a social network, an enterprise wiki, or a corporate bulletin board usually have a lifetime of several weeks. The content on the pages of a typical website has a lifetime that can be estimated in the order of months. Whether it is internal or external, the website must be refreshed after that time.

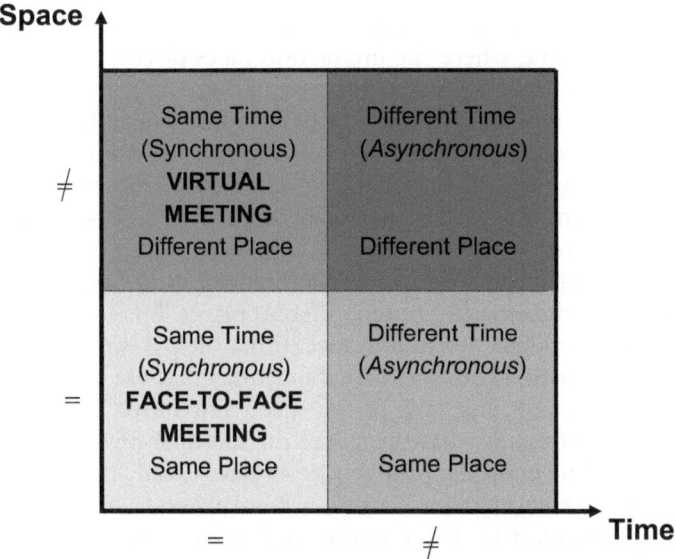

Fig. 5.1   Space / time information sharing

There is, however, some information that needs to be kept and maintained for years. For example, corporate procedures and bylaws, quality certification of internal processes, regulatory procedures, and legal documentation often have lifetimes in the order of several years. They must be stored in safe and long-term storage containers. These containers, like enterprise document management services, tend to have features such as the checking in and checking out of documents, very sophisticated security controls, versioning mechanisms, and both online and offline management.

In many situations the retention period of information is not clear. How long information should be retained is not necessarily known when it is recorded. This is why in many organisations the quick, but expensive, solution is to save everything—just in case.

5.1.2   Information Sharing in Space and Time

One of the most important contributions of ICT is to enable seamless information sharing in space and time (see Fig. 5.1). The matrix in this figure illustrates four different quadrants of information sharing and, therefore, collaboration.

In the bottom left-hand corner, there is the still familiar face-to-face or physical meeting, where the interaction takes place in the same place and at the same time—it is synchronous. In this situation, the human beings who are interacting directly with each other need few or no technologies. A video projector may be sufficient to show images. A shared blackboard or a paper-based flipchart will permit the taking of notes about the meeting. In all the other three quadrants, however, technology plays a central role.

In the top left-hand side quadrant, interactions are occurring at the same time—synchronously—but in different places. It is a virtual meeting. Here ICT is of fundamental importance. It may help to simulate the previous real, live situation through virtual meetings that use audioconferencing or videoconferencing. Yet people need not move from their own locations—attendees just have to agree on the time of the meeting. Of course a virtual meeting can probably never have exactly the same effect as the physical interaction of a real meeting. In many cases, however—as with the tele-consultation or tele-interpretation taking place in hospitals in southern Denmark (Medcom 2012)—it offers a lot of benefits. For example, it reduces travelling time, costs, stress, and pollution.

In the bottom right-hand side quadrant, interactions take place at the same place but at a different time. Here, asynchronous messaging technologies work well: they include email, or even paper notes or sticky notes, since the people are in the same place and can see each other's messages at a later time or date. Imagine a note stuck on a colleague's door if he or she has had to step out for a few minutes or a message displayed on the fridge door in the kitchen at home: 'De-frost the soup for lunch'.

The top right-hand side quadrant, where the meeting takes place in a different (asynchronous) time and a different place, is the example where technologies can offer the biggest contribution in overcoming time and space barriers. Here, emails, virtual rooms, or teamware technologies (software applications that enable people to work as a team even at a distance, also known as 'team groupware') can enable information sharing. Good examples of this asynchronicity are follow-the-sun projects in large international organisations that have satellite offices or branches based in different countries or in different continents.

Two extremely well-known examples of such large organisations are the United Nations (UN) and International Business Machines (IBM) Corporation. The first example is a non-governmental organisation. The UN is an international organisation that was established in 01945 after the Second World War to facilitate cooperation and world peace. Based in New York City, The Hague, Geneva, Vienna, and Nairobi, it also has other offices located throughout the world. It currently includes 193 member states, and it employs more than 50,000 people around the globe. Arabic, Chinese, English, French, Russian, and Spanish are its six official languages. The second example is a commercial organisation. IBM is based in Armonk, New York, in the United States of America, but also in many other countries around the globe. It produces many different products and services, such as computer hardware, software, information technology services, and consulting. Founded in 01911, the firm now employs more than 370,000 people globally.

The evolution in information sharing is heading in directions that are more and more towards one-to-one and many-to-many, even if there are still a lot of broadcast-style (one-to-many) media that are operating such as radio, television, and newspapers. ICT is providing new media where people have the ability to provide feedback or engage in dialogue. The use of these communication or groupware tools is transforming the way people work and collaborate. Since the 01980s, these applications are referred to as Computer-Supported Cooperative Work (Grudin 1994).

These examples are instances of one-to-one communication that, in many cases, is supplanting the old one-to-many broadcasting style of communication. With the existence of networks, there are more and more many-to-many forms of communication available.

5.1.3   How Information Provision Has Developed Over Time

Developments in the electronic era have enabled people to overcome the limits of space, since computer networks are ubiquitous on many parts of the planet. Yet, what about time? One assumes that bits, once stored on a form of support (i.e. a computer), will last forever. Yet, is this correct?

Not all information is, of course, market based. There are some very important exceptions. Some information has a lifetime that is nearly infinity: this is the case of classical masterpieces written by playwrights such as Shakespeare and composed by musicians such as Mozart. Just like the plant seeds located in Svalbard, Norway, and the 10,000-year clock buried in the Texan desert, discussed in Chap. 4, there are indeed some masterpieces that one would like to keep safe forever. There are some kinds of information that people would like to share with future generations. Will these forms of information survive in the long term? Is ICT able to support people in this challenge?

Throughout the last several thousands of years, human beings have traversed several stages—or revolutions—in the communication of information (see Fig. 5.2).

In the beginning, human beings possessed simply the biological support of their minds. It provided them with a kind of wet-ware, a term used since the 01950s to indicate the complex interaction between the physical substrate of the brain and the more abstract concept of mind (Rucker 1988). At some time in the past, human beings had no other available support for their knowledge than their bodies (brain-mind). First, of course, they used gesture, touch, and sound. At a later stage, in the period of pre-history, they used languages to communicate. People told stories and myths to each other—some of which were of great length and importance—handing them on from generation to generation.

A quantum leap took place when people started writing around 3500 BC. Recent history began. This development enabled a jump in the space and time diffusion of information. It was possible, and easier, for information to last longer than a human life. The information support from which human beings benefited no longer took place inside their bodies but outside them, through their use of stones and tablets.

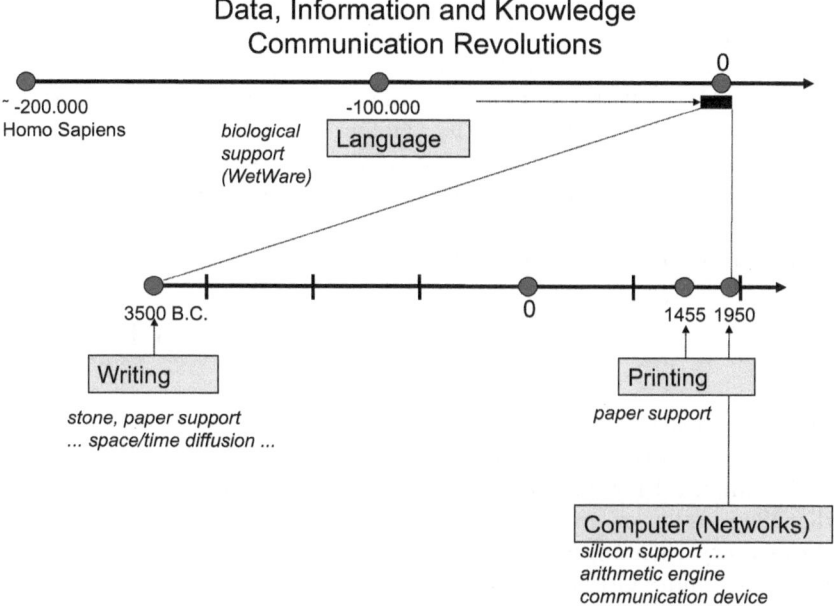

Fig. 5.2   Revolutions in information sharing

Around 01500 AD, human beings started printing information on paper. They could produce thousands of copies of the same information and spread it around in both time and space. Indeed, some commentators have interpreted the techniques used by Martin Luther, when pinning his theses to church doors, as similar to those of contemporary social networking activists (The Economist 2011).

More recently, around 01950 AD, people started to use computers. Support for computing shifted to silicon (i.e. electronic chips). People could now use computers either as an arithmetic engine—which had been their very first use since the mid-nineteenth century—or a communication device, a mode of operation that is predominant today. Computers have both a physical aspect (their hardware) and an intangible side (their software). People can insert information into them (input), store information, process information, and also extract any results or information from them (output). Beyond these four main functions, since 01969, computers have a fifth fundamental property: they are connected to the Internet.

On 5 December 01969, the first computers of the proto-Internet (called ARPAnet) connected four institutions, all in the United States of America: the University of California, Los Angeles; Stanford Research Institute, Menlo Park; the University of California Santa Barbara; and the University of Utah, Salt Lake City (Hafner and Lyon 1998). Since that date, people can transmit and receive information between many different computers: the network of computers that is now called the Internet began here. Users code information in bits. Those bits are stored on hard disks, external devices, or somewhere in the cloud.

5.2   Environmental Limits: Towards the End of Moore's Law?

More powerful and faster silicon chips require a continuous growth in the number of transistors into a single microprocessor.

An empirical law has already been observed for many years. It states that the number of transistors that can be integrated into a chip doubles about every 18 months: it is known as Moore's law. The name refers to Intel co-founder Gordon E. Moore, who described this trend in a 01965 paper (Moore 1965). However, is Moore's law sustainable? Can the processing speed of chips really be doubled every 18 months continuously? Is this enormous power and speed really needed? Are there actually no limits in this direction?

Looking carefully at the chip manufacturing industry, it is clear that these tiny slices of silicon (called chips) lie at the very core of the survival of the contemporary computing era. Yet, they have several drawbacks and a number of negative aspects (Patrignani et al. 2011). There are at least four major issues that pose challenges to the continuity of Moore's law and the survival of chip manufacturing.

First of all, there are the health and safety implications for the people who work inside silicon foundries. These workers face serious chemical hazards due to the use of solvents and gases. Substances like arsine, benzene, hydrogen chloride, tetrachloroethane, and xylene have dramatically negative effects on the human skin, blood, and the central nervous system: they irritate the eyes and can lead to miscarriage late in pregnancy (SJMN 1985).

Second, there are some very serious environmental limits: knowledge of these risks began to emerge around the end of the last decade. A 02010 study which took place at Yale University showed that high-tech products increasingly make use of rare metals of 11 major elements in the 01980s but now use about 60 (two-thirds of the periodic table) (Schmitz and Graedel 2010).

The elements that Schmitz and Graedel (2010) describe are the rare-earths. The 17 rare-earth elements are Scandium (Sc), Yttrium (Y), plus the 15 Lanthanoids: Lanthanum (La), Cerium (Ce), Praseodymium (Pr), Neodymium (Nd), Promethium (Pm), Samarium (Sm), Europium (Eu), Gadolinium (Gd), Terbium (Tb), Dysprosium (Dy), Holmium (Ho), Erbium (Er), Thulium (Tm), Ytterbium (Yb), and Lutetium (Lu). They are dispersed throughout the Earth's crust and are difficult to mine.

Third, perhaps the most politically sensitive challenge is that the resources that are the most economically convenient to mine are concentrated in the territory of a single country, China. Some of the other countries, like Brazil, Russia, India, and South Africa, among those known colloquially as the BRICS, are about to expand their mining potential considerably and this may include mining of the rare-earths.

Fourth is perhaps the most fundamental issue, since it is related to physical limits. According to a recent article in *Nature*, the heat that is generated when more and more circuits are miniaturised and compressed into the same area, and the unavoidable quantum uncertainties of transistors of the dimension of 2–3 nanometres, will make the transistors unreliable: this is the potential twilight of Moore's law. Even more interesting is the reaction of the industry to this situation. In the so-called 'More than Moore' strategy, it was indicated that industry players will investigate the chips needed to support the new generations of sensors, power-management circuits, and other silicon devices needed in mobile environments (Waldrop 2016). It looks like society will have chips that will not necessarily work faster and faster, but they will be designed around the actual power needs of the applications.

These four issues place the sustainability of the entire chip manufacturing industry at risk. Perhaps these risks provide an explanation for why the industry is currently examining the use of other materials for chips, such as organic materials and/or plastics and/or light. Combined with the huge problem of recycling or disposal of the growing mountain of e-waste, if all these arguments are taken into account, they will lead to a serious rethinking both of Moore's law and the entire hardware production chain.

Perhaps therefore more attention needs to be paid to the concept of recyclable-by-design, also known as design-for-recyclability (Chen et al. 1993). This design feature makes a product both easy to repair and to recycle at the end of its life. This recyclability minimises the consumption of resources through reuse and repair and e-waste. It means that the designers already have in mind the environmental challenges at the time of the design of the product or service.

While this is the hardware side of the story, there is also the software aspect.

5.2.1   Computers Are Not Washing Machines

Computers are not standard machines, they are not like washing machines.

Indeed, computers are very special, they are Turing machines. Their name comes from Alan M. Turing (1912–1954), one of the greatest mathematicians of the twentieth century. Turing machines are capable of processing information, which is their mission. The hope is to use this processing power to improve the quality of human life and the environment. Here one can imagine that, in some way, these machines could decrease the entropy generated in producing them. Even better, they could have a payback of positive entropy.

This observation is a reminder of the thought experiment described by James C. Maxwell (1831–1879), known as Maxwell's devil paradox. Maxwell's experiment was one of the first connections made between information, entropy, and energy. The paradox describes a situation in which a little devil could separate fast molecules from slow molecules in a box, by creating a difference in temperature between two sides of the box with no expenditure of energy. As Leo Szilard in 01929 and Leon Brillouin in 01951 both demonstrated, the devil needs to have information about the speed of molecules: simply acquiring information requires an expenditure of energy (Szilard 1929; Brillouin 1951).

Turing machines could perhaps therefore help human beings to decrease the environmental impact created by humanity. They could support people to manage information more effectively. Even more importantly, they could help people to enhance the well-being and well-living of their short lives.

5.3   HUMAN LIMITS: INDIVIDUAL
AND COLLECTIVE BANDWIDTH

Human beings' senses are the communication channels with the real world. Here, the chapter examines both the limitations to individual bandwidth and to more collective forms of bandwidth. It also looks as some activities that human beings are good at.

5.3.1   Individual Bandwidth

In general, human senses have a limited bandwidth in terms of both seeing and hearing. These limits are dynamic: typically people lose these capacities to different extents as they grow older.

Human eyes can see only a small section of the entire electromagnetic spectrum: people cannot see below the minimum frequency of about 400 TeraHertz (this corresponds to a wavelength of 750 nanometres or the colour violet) nor can they see above the maximum frequency of around 790 TeraHertz (which corresponds to the wavelength of 380 nanometres or the colour red). Human eyes experience a maximum sensibility that is centred on the colour yellow (with a frequency of about 540 TeraHertz and a wavelength of 555 nanometres). The collection of colours visible to human beings are those of the rainbow: red, orange, yellow, green, blue, indigo, and violet.

Human ears cannot hear sounds with a frequency that is below the minimum of about 20 Hertz. People cannot hear the subsonic sounds produced by earthquakes or some large church organs; however, some creatures, like elephants, can hear these sounds. Nor can people hear sounds with a frequency of above about 20,000 Hertz. They cannot hear the ultrasounds used by animals like dolphins or whales to communicate among themselves or bats to avoid obstacles in the dark of night.

Human beings are, however, very good at reading. Some people are even extremely fast readers. Yet even fast readers who can read a standard page of about 800 words (or about 4000 characters), and still understand the meaning of what they are reading, cannot reduce the limit beyond two minutes or about 33 characters per second. If code is created for 32 bits to each character, a maximum reading speed can be achieved of about 1000 bits a second, just one Kbit/s.

Digital natives—individuals who as children were raised in the recently digitalised and media-saturated world—use a much more parallel approach to information processing, based on visual interfaces, than do non-digital natives. Non-digital natives grew up with a serial approach to information enforced by the textual interfaces of books that need to be read line by line. Instead, a visual interface can stimulate the human retina, by means, for example, of a typical frame of one Megapixel (1000 × 1000 picture-elements). If each pixel blends three colours, red, green, and blue, each encoded in one byte, then one pixel requires three bytes, or 24 bits, and the frame is made of 24 Megabit (Mbit). By changing the frame 24 times per second, the speed of 24 × 24 Mbit per second or 576 Megabit per second is reached (the perception of movement requires this threshold of a minimum speed of 24 frames per second). With appropriate technologies, this speed can be improved. From 48 frames per second, it will reach about one Gigabit a second. However, human eyes will have difficulties in perceiving this amount of visual information.

Today human beings' minds are immersed with myriad messages, texts, sounds, pictures, and videos. On average, each person produces about one Gigabyte of new data every year. However, by 02011, this figure doubled every three years (UCB 2011). Internet traffic is estimated to reach the 1.6 Zettabyte (10^{21} bytes) threshold by the end of 02018, when one Zettabyte is 10^{21} byte (CVNI 2014).

Since there is a maximum bandwidth of information that human minds can absorb—beyond which information overload occurs—the information flow reaching the human senses has now traversed the physical limits of human minds.

Human beings experience limits in absorbing information and messages. Rather, their minds need time for thinking, meditating, arguing, and questioning. People need to connect with their codes, that is, the collection of background knowledge that they have learnt throughout their lives which is fundamental to their understanding of the surrounding reality. They need to establish and re-establish relationships with their personal histories.

There is the growing risk of a future in which people confuse data, information, knowledge, and, more importantly, wisdom—possessing a set of values that drives them towards an end. Without people's own control of the relevant access codes, interpretation, and with their human senses in an always-on input mode, men, women, and children risk experiencing a situation of total manipulation in which they just become the passive targets of messages. There is a kind of pollution in cyberspace.

5.3.2   Collective Bandwidth

On a more collective and beneficial basis, one can start to design various e-democracy applications, developing ICT tools that enable discussion and the taking of collective decisions. However, can information technology create new models of participation that respect the fundamental values and principles of democracy?

Perhaps one should guarantee a direct participation in cyberspace that would involve the discussion of several decision-making options, and not just involve the counting of opinions as in surveys and e-surveys or e-polls; hearing and being open to other people's opinions, exerting tolerance, and the capability to sustain opinions, and to be able to debate them, in front of other people and not simply to communicate choices that are often restricted to a few options selected from a list.

Fundamental questions arise: can participation be remote, anonymous, and take place without physical contact? Does organising a virtual agora make sense? Other challenges include: How will participatory democracy in Europe develop? How can one expect local councils, national parliaments, and the European Parliament to become? What kind of roles will people, as individuals and groups, play? (Eppler et al. 2012).

Human beings could use technology to improve horizontal communication on a many-to-many basis, and not just be limited to the use of vertical communication as in broadcasting. However, people should also be ready to accept the responsibility of the consequences of their choices and not simply provide anonymous answers to questionnaires.

Democracy needs patience: democracy is necessarily a slow process. Instead, technology risks accelerating decision-making processes dramatically without respect for human timing. For democracy to exist for human beings and between human beings, people need time to reflect, and their minds need time and silence. At times people need to stop receiving inputs in order to influence and change the environment around them.

Advocating freedom of speech for people who live under authoritarian regimes which try to control the Internet is important (RWB 2015). However, there is also the opposite problem of how to provide freedom and space for people in democratic circumstances who are inundated by a surplus of information. People cannot continue to just receive millions of messages (Deleuze 1995). In a different media field, similarly—in his last movie, *La voce della luna*—Federico Fellini advocated that, by creating a little silence, it would be easier to increase understanding (Fellini 1990).

ICT is providing the ultimate tools to disseminate information in both time and space. Yet now people are also starting to observe that there are some limits to this development. These are occurring not only on the environmental side but also in relation to human physiological and neurological constraints on filtering and attending to stimuli.

Slow Tech could provide people with appropriate perspectives for designing and using ICT systems that fit better with environment and human limits, whether individual or collective.

REFERENCES

Bateson, G. (1972). *Steps to an Ecology of Mind*. Chicago: Chicago University Press.

Brillouin, L. (1951). Maxwell's Demon Cannot Operate: Information and Entropy. *Journal of Applied Physics, 22*(3), 334–347. https://doi.org/10.1063/1.1699951. Accessed [9 October 2017].

Chen, R. W., Navin-Chandra, D., & Prinz, F. B. (1993). Product Design for Recyclability: A Cost Benefit Analysis Model and Its Application. In *Proceedings of 1993 IEEE International Symposium on Electronics and the Environment, IEEE Xplore*.

CVNI. (2014). *Cisco Visual Networking Index*. http://www.cisco.com. Accessed [9 October 2017].

Deleuze, G. (1995). *Negotiations*. New York: Columbia University Press.

Eppler, A., Wauters, P., & Whitehouse, D. (2012, March 13). *MEP 2025: Preparing the Future Work Environment for Members of the European Parliament*. European Parliament, Directorate General for Internal Policies, Policy Department D/Department of Budgetary Affairs: Brussels. PE 453.231.

Fellini, F. (1990). La voce della luna, the Last Movie of Fellini, Based on the Novel by E. Cavazzoni (1996). *Il poema dei lunatici*, Feltrinelli.

Grudin, J. (1994). Computer-Supported Cooperative Work: History and Focus. *Computer, 27*(5), 19–26.

Hafner, K., & Lyon, M. (1998). *Where Wizards Stay Up Late: The Origins of the Internet*. New York: Simon & Schuster.

Longo, G. O. (1991, November). Libertà e sicurezza nell'era dell'informatica onnipresente. *Le Scienze, 279*, 106–114.

Medcom. (2012). *Medcom 7*, Project Summary 2010–2011. www.medcom.dk. Accessed [9 October 2017].

Moore, G. (1965). Cramming More Components onto Integrated Circuits. *Electronics Magazine, 38*(8), 82–85.

Patrignani, N., Laaksoharju, M., & Kavathatzopoulos, I. (2011). Challenging the Pursuit of Moore's Law: ICT Sustainability in the Cloud Computing Era. In D. Whitehouse, L. Hilty, N. Patrignani & M. Van Lieshout (Eds.), *Social Accountability and Sustainability in the Information Society: Perspectives on Long-Term Responsibility*, special issue of *Notizie di Politeia, 27*(104).

Rucker, R. (1988). *Wetware*. New York: Avon Books.

RWB. (2015). *Reporters Without Borders, World Press Index Freedom-2015*. http://en.rsf.org. Accessed [9 October 2017].

Schmitz, O. J., & Graedel, T. E. (2010). *The Consumption Conundrum: Driving the Destruction Abroad*. http://e360.yale.edu. Accessed [9 October 2017].

SJMN. (1985, January 17). High Birth Defects Rate in Spill Area. *San Jose Mercury News*.

Szilard, L. (1929). Über die Entropieverminderung in einem thermodynamischen System bei Eingriffen intelligenter Wesen (On the Reduction of Entropy in a Thermodynamic System by the Intervention of Intelligent Beings). *Zeitschrift für Physik, 53*, 840–856.

The Economist. (2011, December 17). *Social Media in the 16th Century. How Luther Went Viral.* http://www.economist.com/node/21541719. Accessed [9 October 2017].

UCB. (2011). *How Much Info—2003.* School of Information and Management Systems, University of California at Berkeley. http://www2.sims.berkeley.edu/research/projects/how-much-info-2003/. Accessed [9 October 2017].

Waldrop, M. M. (2016, February 11). The Chips Are Down for Moore's Law. *Nature, 530*, 144–147. https://doi.org/10.1038/530144a.

CHAPTER 6

The Beginning of a New Renaissance in ICT

Abstract This chapter looks back in history as well as forward, in terms of where computing is heading. The chapter concentrates on the co-shaping of society and technology. To portray this co-shaping clearly, there is an exploration of the history of computing from its early days, throughout the Second World War, to three distinct periods: host computing, personal computing, and cloud computing. While a preoccupation with computer ethics emerged in the late 01950s, in 02017 the thrust is towards a more proactive form of computer ethics. Indeed, Slow Tech forms a part of this general trend. It also acts as a means to introduce a more holistic view of ICT systems and Complex Socio-Technical Systems. In its conclusion, this chapter explores a new direction in thinking, which facilitates a longer-term view of ICT: it is the concept of responsible research and innovation. The ideas are captured in a simple model. This chapter provides thoughtful insights into the contemporary development of Slow Tech.

Keywords Computer ethics • Computing • Co-shaping • Environment • ICT • Innovation • Model • Policy vacuum • Proactive • Renaissance • Responsible research and innovation • Society • Socio-technical systems • Systems theory

© The Author(s) 2018   55
N. Patrignani, D. Whitehouse, *Slow Tech and ICT*,
https://doi.org/10.1007/978-3-319-68944-9_6

This chapter starts from the premise that it is now time to begin to reframe how we think about ICT and its implications in relation to responsibility, sustainability, and ethics. It is important to extend the scope of reflections beyond hardware and the software to include the effects of ICT development and adoption on the society and the environment.

Just as the Italian Renaissance launched an era that placed human beings centre stage, at the start of the twenty-first century, it is similarly time to introduce a new way of looking at ICT which is increasingly humanity-centric in its thinking.

A new Renaissance in ICT is proposed that builds on new awareness and attitudes. It will aim to bring to the fore the tenets of General (and Complex) Systems Theory, and Systemic Design (Von Bertalanffy 1950; Checkland 1981), the findings of *The Limits to Growth* report (Meadows et al. 1972), and the principles of responsible research and innovation in ICT (RRI 2015). As a result, it will situate human beings—and their technologies—in the context of an ecosystem far larger than just that of the human species.

This chapter provides a brief look at how ICT has evolved to a point where it is now able to store, retrieve, process, and transmit information faster than humans can absorb and make sense of it. Human beings continue to embed these technologies more deeply and more extensively in their physical surroundings and their daily activities. This state of affairs ought to raise concerns. Not satisfied with simply pointing out this issue, we want now to introduce some ideas that will help us to reflect more broadly and more deeply on the continued, unquestioning development of ICT, to elucidate all the costs and benefits of such a course and to reassert some control of the speed and direction of this process.

This chapter examines how attitudes towards computers have altered throughout the last 70 years and, as a result, how computer ethics have changed. It launches a longer-term view of ICT and considers sustainability far into the future. It provides a bridge towards the proposal for a positive view of ICT—the opportunity to design, develop, and use Slow Tech, a good, clean, and fair form of ICT. This indeed provides a responsible, sustainable, and ethical approach to technologies.

6.1   Co-shaping of Technology and Society

Taking a rapid overview of the history of computing, it is easy to recognise that changes have taken place in recent times in the relationships between technology and society. This is the co-shaping in action to which Deborah

Johnson refers when she invites people to look at computer-based systems as socio-technical systems (Johnson 1985, 2009). We explore this co-shaping process by looking at five different periods in the evolution of computing: the initial early days, computer use during the Second World War, experiences with host computing, personal computing, and now the growth in cloud computing.

6.1.1   The Early Days of Computing

Even the word computer was shaped by the development of society in the nineteenth century. The word's Latin origin lies in the adjective and the verb *cum-putare* ('together-cut' or 'compare and extract a result'). The expression was used through the seventeenth and eighteenth centuries to designate a person: a computer was, for example, the person in charge of calculating the astronomical tables used by sailors during ships' navigation. In the nineteenth century, the term shifted its application from human beings to the design and creation of a machine that could be used to undertake many of the same calculations. However, the term as it relates to a person as a computer continued well into the mid-part of the twentieth century, particularly with reference to female employees (Grier 2007). So too, with the increasing personalisation of computing and the quantification of the self; its growing intimacy and nearness to—even internalisation in—human beings; and its carriage in and on more organic materials, there is every possibility that the notion of the human being as computer will once again re-emerge and be strengthened.

Naval navigation was of huge importance in the nineteenth century. Society shaped the computing technology needed to facilitate better global travel by encouraging the use of more precise navigation tables. Due to the complexity of navigational calculations, there were a lot of mistakes in early astronomical tables.

In 01822, Charles Babbage proposed the introduction of a technology that could help to overcome these errors (Flood et al. 2011); it was described by Luigi Menabrea and Ada Byron. In 01843, the Analytical Engine was built. Society also shaped technology by embedding values related to the context of the Industrial Revolution. Babbage investigated the organisation of industrial production and the possibilities offered by a general purpose machine, the Analytical Engine—complete with its sequences, branches, and loops—that was programmable through punch cards (Babbage 1832).

At the start of the nineteenth century, the diffusion of technology into society was still relatively low, and the understanding of the impact of technology on society was largely neither understood nor generally studied. Probably the most well known of the events in response to the introduction of technology into the factory setting was the Luddite rebellion, which took place somewhat over 200 years ago (Bailey 1998). With the growth of the working class during the nineteenth century, people started to question the changes that mechanisation was having on their labour rather than the effect of the technologies on themselves.

The Analytical Engine was the first theoretical model of a computer. It remained at the prototype stage and was never produced. However, people still had to wait for an entire other century for the real innovation that was the computer to reach into the whole of society.

This development was based on the intellectual work of Alan Turing (Turing 1937) when he wrote his seminal, but still theoretical, paper. At a similar period, Alonzo Church described the lambda calculus (Church 1932). Considered as one of the fathers of computer science, Turing's studies of cryptography contributed to the defeat of the Nazis at the end of the Second World War. Turing lent his name to three developments of importance: the Turing Machine (the first theoretical model of a computer), the Turing Test (the criterion for determining the difference between a machine and a human being), and the Turing Award of the Association for Computing Machinery (the highest honour in computer science).

6.1.2   Computing Use in Wartime

The most important push towards the physical realisation of computers was the Manhattan Project. It led to the creation and use of the atomic bomb during the Second World War in the twentieth century. In 01943 John Von Neumann obtained the necessary budget and resources to develop the computer architecture that is still in use today; he called his work the EDVAC Report (Von Neumann 1945). That architecture consisted of memory containing data and instructions and a central processing unit that extracted and executed instructions from the memory. Society shaped technology by investing immense resources in the creation of the atomic bomb. As a side effect, this venture made possible the production of computers that used concrete electronic components.

6.1.3   Host Computing

A few years later ICT started to shape wider society: computer applications moved from the military to commercial. Contributions to computing developments were made by Norbert Wiener and Claude Shannon in 01948. The first commercial computers became available. In 01949 there was the Binary Automatic Computer (BINAC), and, in 01951, there was the Universal Automatic Computer I (UNIVAC I) (Johnson 2006).

The year 01950 can be cited as the beginning of the Information Revolution. Now, human beings could use not just time and energy for their activities but also a new resource: information. The importance of information for human beings can be seen in the words of Norbert Wiener: 'The needs and the complexity of modern life make greater demands on this process of information than ever before ... To live effectively is to live with adequate information' (Wiener 1954: 17–18).

With their commercial availability, the computers that had first existed only in research and military laboratories began to enter offices and manufacturing plants. Various technological leaps occurred, such as the development of transistors in 01947, integrated circuits in 01958, and microprocessors in 01971. Nevertheless, the basic architecture of the computer continued to remain quite similar to the original design.

This first age of computer use in society is characterised by the centralised architecture of mainframes (known as host computing) from the 01950s. The impact of computers on society increased during this phase. However, the societal changes occurring were not yet as impressive as they seem to have been later on and are, indeed, today. Host computers' presence was limited to offices and large organisations.

6.1.4   Personal Computing

It was around the 01960s that societal tensions triggered another age of computing: the decentralised and peer-to-peer network architectures of personal computers (Wall Street Journal 1965) and the Internet (Cerf and Kahn 1974). People were pressing for more individual freedoms. Towards the middle and end of this decade, protests against centralised and authoritarian organisations multiplied.

Personal computing started an era of pervasiveness of ICT into society: computers were used not only in offices and manufacturing plants, they were also introduced into people's homes and personal lives (Olivetti 1965;

IBM 1981). The growth of the technology of the personal computer—as well as the lack of single point of control and peer-to-peer architecture of the Internet—began at precisely the same time period that people's empowerment became a major preoccupation in societal terms (Bradley 2017).

A decade or so later, in January 01984, one symbol of this increased desire for freedom—manifested in ICT—was the smart marketing idea celebrated in the 60-second advertisement made by Apple. The Macintosh personal computer was presented as a liberating technology in the struggle against Big Brother: its launch date and the publicity surrounding it were a clear reminder of the threat implicit in the title of George Orwell's dystopian novel (Dougherty 1984).

The personal computer incorporated all the main functions of a computer (e.g. input, memory, a central processing unit, and output) into a single, personal device. It was the best example—until that date—of a technology, ICT, that provided people with autonomy. The contrasts between autonomy and what academics call heteronomy in ICT, that is, actions influenced by forces outside the human being, are explored in this chapter (Patrignani and Kavathatzopoulos 2013).

6.1.5   Cloud Computing

With cloud computing today, in computing terms there is a 'back to the future'-style return to centralised architectures. The autonomy implicit in the age of personal computers is being replaced by a form of heteronomy, in which human beings are no longer computationally independent. Instead, the devices that they use consist of only input and output mechanisms. The core functions of the machine, its memory and processing capability, are now located in the cloud (Patrignani and Kavathatzopoulos 2013: 378).

For many individuals, small companies, and indeed many public sector corporations, the availability of applications and data spaces in the cloud is very attractive. People are becoming increasingly familiar with the notion of using resources that they do not themselves own, as they are also in many other fields of the sharing economy (Bradley 2015; Hamari et al. 2015). With cloud computing, companies can use ICT as a service instead of installing storage and processing machinery on their organisations' sites.

The potential costs to individuals of the convenience of these cloud services are that data in the cloud is collected and used in big data repositories for analytics, data mining, and profiling applications and can thus pose security and privacy concerns. There is currently a large growth in anxiety about many security and privacy challenges. As a result of threats to security, people will have to become more familiar with strong authentication and improved security levels. Researchers are increasingly investigating the social consequences of cloud computing adoption on a large scale (IFIP 2015; Mosco 2014). More widely, they investigate the different paths followed by ICT in reference to the varying historical contexts in which its development took place.

Many interesting questions arise when cloud computing is considered as a socio-technical system, even more so when it is recognised that this system lies within a larger ecosystem: What kind of co-shaping is taking place between cloud computing and society? Why is the type of society that exists at the beginning of the twenty-first century leading towards a cloud computing-based ICT? What kind of society will be shaped by this new centralised architecture of ICT? Will this new direction in technology lead to a more consolidated society? Will people lose the status of digital citizens to become instead digital consumers (Patrignani and Kavathatzopoulos 2013)?

Cloud computing, as a combination of distributed storage/computing and extensive telecommunications networks accessible through browser software, can be seen as a platform for 'disappearing computing', a co-funded European Commission proactive initiative which formed part of the Future and Emerging Technologies activity with its mission to see '… how information technology can be diffused into everyday objects and settings' (DC 1998).

6.2   Towards a Proactive Computer Ethics

The view of complex computer-based systems as socio-technical systems is the key that opens the door to a more positive view of ICT. We propose to enter this door with a Slow Tech compass.

In the field of ICT, today's era is still one in which technology push dominates. Society is losing out under the pressure of the ICT companies. A consumer-oriented view of technology remains mainstream, and it is perhaps growing in certain fields such as the health domain.

People need to find a way to look at what is happening by lighting up the corners in the ICT world, illuminating it from several different angles so that nothing remains in the shadows. This will enable society to decide whether the direction that the ICT industry is currently taking is indeed the desired, and appropriate, one.

There is increasingly a need for new ICT designers who apply, system by system, an ethical framework that accommodates all stakeholders, who are aware of the effects of ICT on people individually and collectively and on the planet. Computer professionals need to recognise that these systems are, in reality, soft and messy and often have no hard borders. However, there are approaches, methods, and tools that will enable such professionals to elucidate the visible and invisible workings and effects of systems and to understand them better as a whole.

Changes are underway: the first signals of a new awareness are appearing. It is now time to question how ICT is used from the societal point of view. This questioning arises as a result of certain forces. People are experiencing a perceived lack of time due to information overload; there is a tsunami of data from sensors to the Internet of Things; there is a growing awareness of the e-waste challenge; and it is even possible to hear and read about the often terrible physical conditions of workers employed in ICT manufacturing plants. What kind of actions can be envisaged to counteract the current dominance, at times negative, of ICT in people's lives?

Ecology and environmentalism can be taken as specific examples of systems that should be much more human-centred. In environmentalism, parallels can be drawn from other fields, such as the automotive industry and space travel. In car manufacturing, there is a growing awareness of the impact of people's mobility on the environment and on climate change. This insight is encouraging a new concept of transportation, based on a wide range of vehicles: electric cars, cars designed by using a recyclable-by-design approach, and car-sharing business models that are called car-as-a-service (Bishof 2014). With cars, this co-shaping is taking a direction in which the technology is becoming more society- and environment-friendly.

Environmentalism provides certain pointers to what may occur in the ICT field. Sometimes complexity is affected by small events which may eventually have massive implications. In the past, a number of events have occurred after which people's relationship with the planet has been modified forever. Two examples spring to mind. For some commentators, for example, the simple 01968 photograph of the finite spaceship earth, taken from space, changed human consciousness permanently about the limits of planetary resources (NASA 1968). For others, it was the publication of the Club of Rome report that, for the first time in human history, introduced the concept of the limits to growth on a finite planet (Meadows et al. 1972).

This overview of ICT and its comparison with two specific environmental examples permit a focus on the growth of an ethics of computing: these are ways of examining how ICT can be designed, produced, and used in more ethical ways. There is a shift towards a more proactive form of computer ethics.

As a result, it is feasible to examine three more detailed eras of computing ethics: what can be called its dawn; the development of a policy vacuum; and the contemporary expansion of a more proactive form of ethics. Essentially, the three eras move through various stages: a raising of critical issues; the actual naming and later the definition of computer ethics; and the shift towards, first, a socio-technical and, second, a more systemic perspective on computer ethics. This stage model of computer ethics highlights the need to formulate a more long-term view of ICT and society. It permits the transposition into a Slow Tech approach to ICT.

6.2.1   *The Dawn of Computer Ethics*

Questioning of the ethics of information technology approaches started at a relatively early stage of ICT development. Besides being the initiator of cybernetics, one of the early leaders of the computer era, Norbert Wiener, a Professor at Massachusetts Institute of Technology, is also considered to be the founder of what today is called computer ethics (Bynum 1999). From the start of his work, Wiener invited the scientific community and society to analyse the social and ethical impacts of computing technologies (Wiener 1954, 1961).

It was not, however, until the 01960s that the term ethics, in association with ICT, can be seen concretely in literature for the first time. As Donn Parker wrote: 'It seemed that when people entered the computer center they left their ethics at the door' (Parker 1968: 198). With the publication of this paper, debates about whether, from the perspective of moral guidance, there are different behaviours in the real and virtual worlds arise (Parker 1968). In 01978, it was Walter Maner who first introduced in writing the term computer ethics. In a starter kit on the subject, Maner wrote about the need for a new applied ethics. Computers, he argued, generate wholly new ethical problems that would not have existed if computers had not been invented (Maner 1978). Calling the proposed new field computer ethics, he described an ethics that studies ethical problems that are aggravated, transformed, or created by computer technology (Maner 1978).

At the same era, Joseph Weizenbaum, a colleague of Wiener's at Massachusetts Institute of Technology, one of the inspirers (pneumatophores) of this book, wrote *Computer Power and Human Reason: From Judgment to Calculation*, a thoughtful volume that emphasised the importance of reflecting on the use of information technology and the risk of delegating fundamental choices to machines (Weizenbaum 1976).

Around the same time period, although in a more environmentally related setting, originally writing in 1979, Hans Jonas underlined the importance of making appropriate technological choices with regard to the planet and future generations (Jonas 1985). Jonas insisted that human survival depends on the efforts of humankind to care for the planet and its future. He also wrote that both currently living beings and future generations should be included in the ethical debate. In particular, this was due to the immense power that technology has placed in human hands and its potential impact on nature (Jonas 1985). The further development of ICT has increased this power even more.

6.2.2   A Policy Vacuum

This new branch of applied ethics was especially needed in the late 01970s and early 01980s because of the beginning of the introduction of computers into people's lives, workplaces, homes, and schools. In 01985, the first actual definition of computer ethics was provided by James H. Moor (Moor 1985). He drew to readers' attention the idea that a policy vacuum existed in the field of ICT and that thus a central task of computer ethics was to formulate such policies (Moor 1985: 1). Of course, the main challenge remains the need to define the process to determine what people should do as a result.

Moor's work assumes that there is always a delay between the time at which ICT is developed and its use in society. Often ICT evolves so fast that people possess no immediately relevant conceptual background to deal with the technology. They have very little time available to them in policy decision-making terms to develop a framework for deciding between right and wrong ways to behave, whatever the ICT scenario.

Computer ethics played an important role at this stage of evolution in ICT development: as an area of philosophy, it helped to provide policies that could lay down ethical behaviour in relation to ICT. In this sense, this computer ethics was a compensatory or reactive form of ethics. In the mid-01980s, there was no questioning yet of the actual ethical design of computer systems.

Today, once again, there is a similar gap in terms of a policy vacuum. The vacuum arises as a result of computer use in practice. Technology is running ahead at a fast pace, and the people's reactions lag behind. People are experiencing a consequent lack of guidelines and ethical guidance. Society is therefore beginning to question the contemporary evolution, speed, and direction of ICT.

6.2.3   Further Shifts in Computer Ethics

In the more than 30 years that have followed Moor's work, a maturation process has taken place in computer ethics thanks to the work of a number of researchers. The focus on computer ethics has undergone a real growth in attention, thinking, and development.

Among the most important people in the community of ethics researchers are Deborah Johnson with her concept of computers as socio-technical systems (Johnson 1985, 2009), Batya Friedman with her focus on value-sensitive design (Friedman 1996), and Helen Nissenbaum with her work on societal, ethical, and political dimensions of information technology and digital media (Nissenbaum 1998). The latest contribution to this line of thinking is the proposal made for ethical IT innovation by Sarah Spiekermann (2015).

The socio-technical systems perspective is particularly important in understanding the evolutionary path of ICT. Deborah Johnson wrote about three science, technology, and society recommendations, which for her form the basis of 'socio-technical computer ethics' (Johnson 1985, 2009). She described these recommendations as being ones that would enable the rejection of technological determinism and the notion of technology as either material object(s) or as neutral. Her main message was, instead, in favour of a co-shaping of both society and technology (Johnson 1985, 2009: 13–18).

The work of Friedman and Nissenbaum, known as value-sensitive design, was crucial for taking human values into account in the design of information systems and human-computer interaction (Friedman 1996; Nissenbaum 1998).

Spiekermann's systems design approach (Spiekermann 2015) provides contemporary insights into a value-based approach applicable to the technologies developing in this decade.

With the launch of these research directions, the mid-01990s was probably the first period of interdisciplinarity to be seen in computer ethics: ethicists, engineers, and designers began to talk to each other about ICT. Indeed, with the expansion of an interest in human-computer interaction, computer scientists themselves were experiencing working in collaborative, interdisciplinary teams (Harrison et al. 1994).

Today, in contrast, designers of ICT systems often start their careers by working in interdisciplinary teams that involve experts from fields such as anthropology, human-computer interaction, philosophy, and sociology (EIT 2015). This cross-disciplinary collaboration opens up a fantastic opportunity to question technology evolution.

As Deborah Johnson (Johnson 1985) argued, technology does not just consist of artefacts but—rather—incorporates values from society; technology and society co-shape each other, and ICT systems are socio-technical systems. As a result, engineers and designers bear a social responsibility and are accountable for their design choices. The good news, implicit in this argument, is that ICT evolution and direction can be changed. Human beings have choices.

Thus, the socio-technical systems concept provided an intellectual and analytical turning point: it offered—and offers still—the possibility to forge a more proactive computer ethics.

6.2.4   Proactive Computer Ethics

Among the vital ethical questions to be asked could now become: What values drive the ICT evolution? What kind of norms, explicit or implicit, are designers incorporating into technology? These questions represent a groundbreaking shift in the direction of evolution towards a more proactive computer ethics. Society can no longer simply try to cope with the pace and speed of ICT evolution, it must instead shape technology. This is the start of a new way of looking at technology. The computer revolution is not inescapable in its present form. Even more importantly, what is the process to be used to determine these values? Can a Slow Tech approach provide more time for this process?

The idea that software design itself can shape society was taken further by Professor of Law, Lawrence Lessig, at the end of the 01990s when he argued that 'code is law' (Lessig 1999). He thereby implied that both computer codes and legal codes can act as instruments of social control. In his groundbreaking book on cyberspace (Lessig 1999), Lessig introduced a simple model for the regulation of complex scenarios that need governance. The model is based on the four approaches of market, law, norms (education), and architecture (Lessig 1999: 88). Governance through architecture is based on the idea that human reality is increasingly influenced by rules that are embedded in software code: '... the software and hardware that make cyberspace what it is constitute a set of constraints on how you can behave' (Lessig 1999: 89): that is, the technological architectures of computers and networks rule cyberspace and, hence also, people's lives. This simple statement has tremendous implications for society in three areas in particular. They are the need for transparency in the software development process, an accountability on the part of software designers, and a computer professional code of ethics or code of conduct.

A good synthesis of this new view of ICT was provided by the founder of business ethics, Richard De George, who wrote that: 'Computers and Information Technology should help and serve people and society. Where they do not, they should not be passively accepted' (De George 2003: ix).

As in any area of human activity, the co-shaping process sees human values embodied in the manufacture, distribution, and use of ICT. At the same time, the technology together with its immediate effects and the expectations it creates alters these values in ways that are not always immediately perceived. The risk is that people collectively allow technological determinism to control where in society and in their lives ICT becomes embedded. They then discover that they are unable—or unwilling—to remove it.

We therefore need to find a way to look at what is happening, illuminating the developments from several different angles so that nothing remains in the shadows. We need to decide whether these actions are ones that we really want to take and whether technology actually does facilitate people's becoming what they really want to be. We need technologists to adopt and apply, system by system, an ethical framework that accommodates all stakeholders and tries to see all the effects of ICT on people, whether as individuals or collectively, and on the planet. There are, however, approaches, methods, and tools that will enable such professionals to elucidate the visible and invisible workings and effects of a system and to understand them better as a whole.

Designers of ICT systems can start a dialogue with society at the design stage and can look at ICT with a human-centred approach, throughout the entire life cycle of technology. For example, Don Gotterbarn and Simon Rogerson presented the possibility of undertaking in-depth investigations and analyses of the risks implicit in the software development process (Gotterbarn and Rogerson 2005).

Today, the mantra that technology is driving the future is beginning to lose ground. Many of the approaches that lie behind a technology push are being questioned. The need to rethink computer ethics by using a less reactive, and more proactive, approach is at the basis of a new way of looking at ICT. Computing innovation needs to be steered more constructively, positively, and responsibly. It is for this reason that the Slow Tech concept can be used as a compass, steering the human quest towards a proactive computer ethics and guiding human choices in the direction of a good, clean, and fair ICT (Patrignani and Whitehouse 2014).

6.3   ICT as Complex Socio-Technical Systems

By looking at ICT as a product of designers' choices, the concept of ICT as socio-technical system(s) opens up the opportunity for a new systemic view of the entire ICT life cycle. Underlining the importance of the inclusion of human beings into the ICT design process implies a more complex form of system design. Designing ICT systems is becoming the design of complex systems. The question arises: is a systemic design of ICT possible?

6.3.1   Systemic Design

System theory has deep roots in the history of science (Von Bertalanffy 1950; Emery 1969; Checkland 1981; Laszlo 1996).

In general, systems theory promotes the awareness of connections in complex systems. It emphasises the importance of looking at the whole system's emergent behaviour rather than concentrating on single components of the system. It conceives the design discipline as a significant bridge between human beings and technologies: it focuses on the interactions among the various system components.

Systemic design takes into account the identities of the components, their relations, the direction of development, and the system evolution. In this field, the goal and the desired direction are the collective well-being of humans and the environment. The interactions among the components are designed by consideration not just of the goal but also of the importance of the respect of the components' identities (Systemic Design 2015). The system is defined by the emerging property of myriad interactions between a large number of elements. By definition, the environment is considered to be multi-stakeholder (Jones 2014). Examples of such multi-stakeholder systems are a healthcare system, a transportation network, or an ICT system.

In the early days of computer history in the 01950s, computers were viewed as just a by-product or a side-effect of a more general discipline, cybernetics, which was in turn considered to be the 'queen' of systems' design (François 1999). Today, with the rediscovery of complex systems theory and network science there is a return to basic fundamental and philosophical principles (Science 2009).

6.3.2   Importance of Complex Systems in Computer Science

Why is it so important to study complex systems? Why have computer science curricula tended to avoid this important connection with the real world? There are likely to be several reasons.

Several scholars have investigated why cybernetics has been dropped from tertiary education. It is possibly due to the deepening specialisation of scientific research (Pruchnic 2013: 8) or to the increasing market pressure for growing numbers of programmers (BLS 2014), when the energies of computer science and engineering students have been concentrated purely on programming. Some signs of a need for a more cross-disciplinary approach are now visible, however: for example, in some countries, since the 01990s, there has been an introduction of computer ethics courses in computing curricula (Tucker 1991). Understanding reality implies the need to face the challenging scenarios posed by complex systems.

The roots of complex systems in computer science go back to the mid-twentieth century. Warren Weaver was one of the more important scientists and mathematicians of the twentieth century, who worked with Claude Shannon on information theory (Shannon and Weaver 1949). Weaver pointed out that science has to face three classes of problems (Zhou et al. 2015: 428). The first is the class of simple systems in which there are few variables that can be easily described by mathematical functions. The second is the class of disorganised complex systems: these systems are typified by a large number of variables. Their unpredictable, microscopic behaviour is very difficult to describe by using only simple mathematical formulae. These systems require the tools of statistics. The third class of organised complex systems contains a large number of variables that interact with each other by exchanging messages and information. They are characterised by their emerging properties. The human body and networks of cells are very good examples of such systems. These systems cannot be studied by just formulae or statistics. They need new methods that are based, for example, on simulations and observations of historical behaviour. Their sophistication and complexity means that they need to be studied by interdisciplinary teams. According to Weaver, science needed over the next half century to handle the problems posed by organised complexity (Weaver 1948: 540).

Designing computer-based systems, complex human-computer interactions, and taking a more holistic view of ICT, implies using a systemic design approach. This systemic conception is now one of the most challenging and interesting areas of science and research. Complex systems, network science, and big data analysis are all new areas of this systems thinking. Computer professionals and scholars are learning from the leading frontiers of science that it is of fundamental importance to include human beings and the dimensions of time and history into the systems they are designing. Designing computer-based systems is a kind of complex systems engineering challenge. Like many other professions, ICT people are involved in applying these views of life and understanding their implications (Capra and Luisi 2014).

Looking into the future, there is the need to recognise that complex socio-technical systems are not constant over time nor are the consequences of changes due to such systems, such as the continued evolution of ICT, entirely predictable.

6.4   Innovation

Historically, innovation is one of the landmarks of human society. It is not an option, rather in the words of Heraclitus, 'everything changes and nothing stands still' (Diogenes Laertius 1925; Johnson 2010).

Research can change society through innovation. A huge idea generation engine acts as the trigger for innovation, and the best research centres and innovation communities nurture the free flow of ideas. Through innovation, a new way of doing things can be brought about, and new products and services can be made available to everybody. But where do innovations and new ideas come from?

There are two common ways of innovating. The first approach is the classical problem-solving approach; the second is more visionary. First, with problem-solving, there is a perceived social or environmental challenge that needs a solution. This situation can be termed, as in the common parlance of the English proverb, 'Necessity is the Mother of Invention'. This is also a typical engineering approach. Yet today's novelty is that, since both the problem and the solution are social constructions, they should also be considered as socially desirable. Second, some needs or problems can be identified as belonging to a next future. A future condition can be intercepted by the creation of an appropriate solution. This approach can be called 'The Future Exists' (Gold 1998). More precisely, if people are able to intercept the particular problem at the right moment in the future, then the resulting solution, product, or service will provide social and economic benefits. This approach produces a challenge for human beings in terms of their capabilities to foresee the future. As suggested by the Long Now Foundation, humans need to include the planet's limits and the capacity for a larger view of humanity into the far distant future (TLNF 2015).

6.4.1   Innovation, Ethics, and Responsible Research and Innovation

At the beginning of the twenty-first century, social and environmental challenges are becoming central challenges for innovation. Innovation on its own is no longer sufficient. A newer, emerging approach to research, science, and technology is attracting a growing interest: it is called responsible research and innovation. The principles of responsible innovation could help in taking appropriate directions in the search for a road ahead. In particular, in the field of ICT, these responsible approaches to innovation are maturing in the research and policy maker communities.

Yet, what precisely are the ideas worth pursuing? What are the innovations that one would consider responsible? The formulation by the European Commission of Responsible Research and Innovation (RRI) in its Horizon 2020 programme is based on the following five 'pillars':

- Engagement ('Choose together'),
- Gender Equality ('Unlock the full potential'),
- Science Education ('Creative learning fresh ideas'),
- Ethics ('Do the right "think" and do it right'),
- Open Access ('Share results to advance') (RRI 2015).

It is worthwhile underlining the fourth, 'ethics', point in which—as philosophers suggest—it is recommended to think in the right way.

Several recent research projects in the field of RRI have produced interesting results. One of the most important is the framework for responsible research and innovation in ICT. It suggests four steps for the analysis of a research project: Anticipate (forecast the outcomes), Reflect (can it be done differently?), Engage (include all stakeholders), and Act (what changes should be implemented) (FRIICT 2014). Other researchers propose a more general framework for addressing the problem of democratic governance in science and innovation and their related social and ethical concerns (Stilgoe et al. 2013).

Innovation means mapping an idea onto a product or a service. If an idea is not translated into something that customers can buy and use, then maybe it has produced interesting material but simply as documentation for a paper or a book.

One of the most powerful schemas describing innovation (taking an invention to the market) is the one based on four waves: creativity, feasibility, prototyping, and engineering (Bivee 2014). Bivee's scheme indicates that the first output from the creativity step (maybe a description of a brilliant idea) must pass through the filters of the feasibility step (is the idea implementable?). The feasibility study must be implemented in a prototype: the proof of concept that demonstrates that the idea can become a real product or service. Finally, the prototype must be analysed in an engineering phase: this will demonstrate that it is possible to find suppliers for all the components needed, to integrate them so as to build the product, maintain and repair it, and provide it to customers. Each of these four stages that contribute to taking an innovative idea to market has its own stakeholders' network: the actors in the creativity phase, for example, are usually different from the ones in the engineering phase.

In principle, at each of Bivee's four stages, the RRI framework should be applied with its focus on engagement, reflection, inclusion, anticipation, and action—that is, engage with all stakeholders; reflect with involved stakeholders on all dimensions of innovation; include considerations on all dimensions of innovation in research processes; anticipate social, environmental, and ethical effects; act and respond to stakeholders' feedback by changing directions in research processes. These steps can be reinterpreted in a more compact model that shows the continuous dynamic adjustments needed to embed ethical reflections into the procedure. While RRI is an important theory, it could prove to be even more important in practice. It is possible to model the way in which RRI can work in real life.

It is important to underline that all the above is a new design process. It needs time to be applied. Yet, even more importantly, it requires new skills from engineers and others involved in the production and design process. This, of course, implies either new curricula or new content in curricula (Hammons et al. 2017), new design tools, and the ability to review upcoming project plans.

6.5   A MODEL FOR RESPONSIBLE RESEARCH
AND INNOVATION

The process of knowledge production matters in research. Innovation is important when a creative idea is translated into either a feasibility study or a prototyping check, and an engineering activity is transformed into an actual product or service by a real-world company.

An important step towards a more long-term and sustainable approach to ICT design is included in the definition of RRI introduced by the European Commission in its Horizon 2020 framework (RRI 2015). It is defined as an approach that 'anticipates and assesses potential implications and societal expectations, … with the aim to foster the design of inclusive and sustainable research and innovation' (RRI 2015: 1).

In the field of ICT, this implies that researchers, industry, users, citizens, and policy makers need to collaborate in order to achieve the values, needs, and expectations of society with an interdisciplinary approach. Thoughtful design is of primary importance in ICT-related research and innovation and project design. An appropriate mix of intellectual and scientific disciplines is also key in any project team (RRI-ICT 2015: 1).

Over the past five years or so, researchers have made efforts to map RRI guidelines into more practical actions in ICT (Stilgoe et al. 2013; FRIICT 2014; Stahl et al. 2014). Some interesting frameworks have been developed. A strong connection is emerging between the evolution in computer ethics, that is, the view of computers as socio-technical systems, and the importance of designing complex systems that involve all stakeholders. Indeed, RRI applied to ICT contributes not only to the evolution of computer ethics but also to a broader approach to the governance of science, technology, and innovation (Stahl et al. 2014: 810).

One of the main difficulties that emerges when designing a complex ICT system is the fact that ethical issues are often dealt with very late in the timeline of any initiative. This delay in action means that it can be too late in the process to address the ethical issues seriously. Organisationally it would be much easier to address these ethical concerns at the very beginning of a project when there is still space and opportunity for changes to be made in the project definition. It can, however, be very difficult to forecast social and ethical problems at such a start-up phase. This challenge is familiar to researchers in the field of technology and society relationships: it is known as the Collingridge Dilemma (Collingridge 1981).

As a result of the Collingridge Dilemma, a number of questions arise: When beginning an ICT initiative—whether it is a research-based or deployment-based—how is it possible to address social and ethical problems in a more flexible and dynamic way? How is it feasible to include all stakeholders in the knowledge production and innovation processes? How is it possible to embed ethics into procedures?

A possible response to these complex design questions is provided by a synthetic closed loop approach (see Fig. 6.1). This model may be useful when designing a dynamic RRI-related system. It has two main purposes: first, it addresses the Collingridge Dilemma and, second, it includes the most recent contributions of contemporary researchers in the fields of innovation and responsible research in ICT (Stilgoe et al. 2013; FRIICT 2014; Stahl et al. 2014).

The proposed scheme is circular. It is a continuous feedback process that is able to correct promptly the direction of an initiative or project by incorporating feedback from stakeholders involved in the process. Since the model is based on a timely response to feedback expressed by stakeholders, it minimises the risk implicit in the Collingridge Dilemma of the

A Model for Responsible Research & Innovation

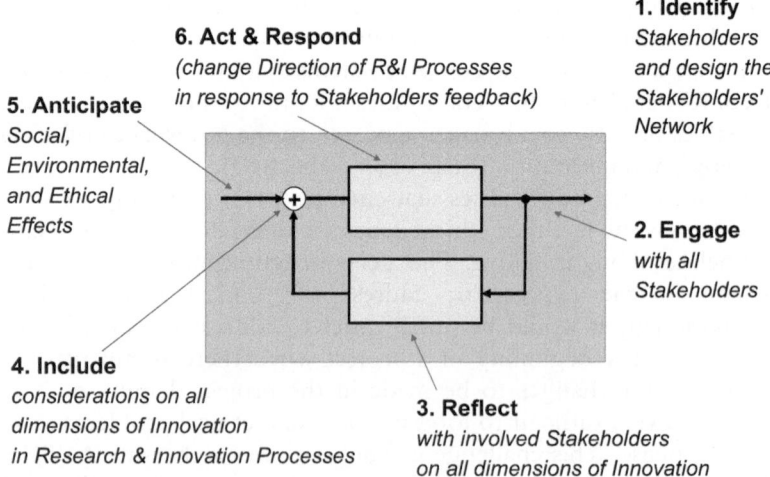

6. Act & Respond
*(change Direction of R&I Processes
in response to Stakeholders feedback)*

5. Anticipate
*Social,
Environmental,
and Ethical
Effects*

1. Identify
*Stakeholders
and design the
Stakeholders'
Network*

2. Engage
*with all
Stakeholders*

4. Include
*considerations on all
dimensions of Innovation
in Research & Innovation Processes*

3. Reflect
*with involved Stakeholders
on all dimensions of Innovation*

Fig. 6.1   A model for responsible research and innovation

difficulties accumulated by the experience of providing a delayed response to ethical issues.

This RRI model is based on the following six steps that should be followed before starting a project or initiative. They start from the right-hand side and proceed clockwise:

1. **Identify**: Identify all stakeholders involved and design the most complete stakeholders' network that best describes in detail the scenario of the initiative.
2. **Engage**: Engage with all relevant stakeholders.
3. **Reflect**: Reflect with the stakeholders involved on all the dimensions of innovation.
4. **Include**: Incorporate the considerations on all the dimensions of innovation in the responsible research and innovation process.
5. **Anticipate**: Anticipate the social, environmental, and ethical effects of the initiative. (This action provides the input to what should be the actual beginning of the research/innovation initiative.)
6. **Act and respond**: Adjust the direction of the responsible research and innovation in response to stakeholders' feedback. As a result, the research/innovation processes are updated dynamically, by demonstrating that the feedback from the stakeholders has affected the actual direction of the initiative. As stated earlier, this process can be described as a dynamic feedback loop.

Note: A stakeholder's network is a graphic display of nodes and arcs, in which the nodes represent the stakeholders, and the arcs the relationships among them (Patrignani and Kavathatzopoulos 2013).

This proposed scheme provides the introduction to a more long-term and sustainable approach to ICT design. It identifies that ICT design takes time and patience, because it involves stakeholders, examines all the relevant complexities, and is flexible enough to prepare for change. It enables industrialists, activists, and researchers to contemplate a longer-term view of innovation.

The Slow Tech approach can enter the scene at precisely the point when RRI embraces socio-technical systems or complex general systems principles and is applied specifically to an ICT scenario.

For all the stakeholders involved, the three main dimensions of Slow Tech can be examined with regard to themselves and their relationships with each other. Is the technology good? That is, is it human-centred, socially desirable? Is it clean? Environmentally sustainable, recyclable-by-design? Is it fair? Therefore, is it respectful of human rights throughout its entire supply chain?

This Slow Tech approach is all about skills and processes that will enable the development of better products and services: essentially, goods, equipment, and behaviour that will manifest a responsible, sustainable, and ethical approach to ICT.

REFERENCES

Babbage, C. (1832). *On the Economy of Machinery and Manufactures.* Pall Mall East: Charles Knight.

Bailey, B. J. (1998). *The Luddite Rebellion.* New York: New York University Press.

Bishof, M. (2014). Automobile-as-a-Service: How Will Your Car Serve You? *Wired.* http://www.wired.com/insights/2014/10/automobile-as-a-service/. Accessed [9 October 2017].

Bivee. (2014). *Business Innovation in Virtual Enterprise Environment.* EU FP7 Project no.285746 on the theme FoF-ICT-2011.7.3—Virtual Factories and Enterprises. http://bivee.eu. Accessed [9 October 2017].

BLS. (2014). Bureau of Labor Statistics. *Occupational Outlook Handbook.* http://www.bls.gov/ooh/computer-and-information-technology/computer-programmers.htm. Accessed [9 October 2017].

Bradley, K. (2015). Open-Source Urbanism: Creating, Multiplying and Managing Urban Commons. *Footprint Delft Architecture Theory Journal, 16, 9*(1), 91–108.

Bradley, G. (2017). *The Good ICT Society. From Theory to Actions.* Abingdon, Oxon: Routledge Focus.

Bynum, T. W. (1999, July 14–16). The Foundation of Computer Ethics. Keynote Address at *The Australian Institute of Computer Ethics Conference 1999 (AICEC99)*, Melbourne.

Capra, F., & Luisi, P. L. (2014). *The Systems View of Life, A Unifying Vision.* Cambridge: Cambridge University Press.

Cerf, V. G., & Kahn, R. E. (1974). A Protocol for Packet Network Intercommunication. *IEEE Transactions on Communications, 22*(5), 637–648.

Checkland, P. (1981). *Systems Thinking, Systems Practice.* Chichester: Wiley.

Church, A. (1932). A Set of Postulates for the Foundation of Logic. *Annals of Mathematics (2nd Series), 33*(2), 346–366.

Collingridge, D. (1981). *The Social Control of Technology.* London: Palgrave Macmillan.

De George, R. T. (2003). *The Ethics of Information Technology and Business.* Hoboken: Blackwell.

Diogenes Laertius. (1925, January 1). Quoted in *Lives of Eminent Philosophers* (R. D. Hicks, Trans.). Loeb Classical Library.

Dougherty, P. H. (1984, June 26). Advertising: Chiat Wins at Cannes for '1984' Apple Spot. *The New York Times.*

EIT. (2015). European Institute of Technology, ICT Labs, Master School of *Human Computer Interaction and Design.* www.masterschool.eitictlabs.eu/ programmes/hcid/. Accessed [9 October 2017].

Emery, F. E. (Ed.). (1969). *Systems Thinking.* Harmondsworth: Penguin Books.

Flood, R., Rice, A., & Wilson, R. (2011). *Mathematics in Victorian Britain.* Oxford: Oxford University Press.

François, C. (1999). Systemics and Cybernetics in a Historical Perspective. *Systems Research and Behavioral Science, 16,* 203–219.

Friedman, B. (1996, November/December). Value Sensitive Design. *Interactions.*

FRIICT. (2014). *Framework for Responsible Research and Innovation in ICT.* responsible-innovation.org.uk. Accessed [9 October 2017].

Gold, R. (1998). *The Seven Pragmas of Innovation.* Presentation given at Palo Alto Research Center on 8 October 1998.

Gotterbarn, D., & Rogerson, S. (2005). Next Generation Software Development: Responsible Risk Analysis Using SoDIS. *Communications of the Association for Information Systems, 15,* 730–750.

Grier, D. A. (2007). *When Computers Were Human.* Princeton: Princeton University Press.

Hamari, J., Huotari, K., & Tolvanen, J. (2015). Gamification and Economics. In S. P. Walz & S. Deterding (Eds.), *The Gameful World: Approaches, Issues, Applications* (pp. 139–161). Cambridge, MA: MIT Press.

Hammons, R. L., Patrignani, N., & Whitehouse, D. (2017). The Slow Tech Journey: An Approach to Teaching Corporate Social Responsibility. *Design & Technology Education: An International Journal.* Manuscript submitted for publication.

Harrison, B., Mantei, M., Beirne, G., & Narine, T. (1994). *Communicating About Communicating: Cross-Disciplinary Design of a Media Space Interface.* http:// www.dgp.toronto.edu/OTP/papers/x.disc.design.ms.interface/X.Disc. Design.MS.Interface.html. Accessed [9 October 2017].

IBM. (1981, August 12). *Personal Computer Announced by IBM.* Press release by Information Systems Division, Entry Systems Business.

IFIP. (2015). *Towards a Human-Centred Cloud Computing: An International Perspective for the Public Interest.* Position paper of the International Federation for Information Processing Domain Committee on Cloud Computing (IFIP-DCCC). Laxenburg: International Federation for Information Processing. Available at http://www.bfia.be/home/towards-a-human-centred-cloud-computing-an-international-perspective-on-the-public-interest. Accessed [9 October 2017].

Johnson, D. G. (1985). *Computer Ethics.* Englewood Cliffs: Prentice-Hall.

Johnson, L. R. (2006). Coming to Grips with Univac. *IEEE Annals of the History of Computing, 28*(2), 32–42. https://doi.org/10.1109/MAHC. 2006.27

Johnson, D. G. (2009). *Computer Ethics* (4th ed., pp. 13–20). Englewood Cliffs: Pearson International Edition, Prentice Hall.

Johnson, S. (2010). *Where Good Ideas Come from: The Natural History of Innovation*. New York: Riverhead Books.

Jonas, H. (1985). *The Imperative of Responsibility: In Search of an Ethics for the Technological Age* (H. Jonas & D. Herr, Trans.). Chicago: University of Chicago Press. Originally published as *Das Prinzip Verantwortung: Versuch einer Ethik fur die technologische Zivilisation* [Frankfurt am Main: Insel Verlag, 1979].

Jones, P. (2014). Systemic Design Principles for Complex Social Systems. In G. Metcalf (Ed.), *Social Systems and Design, Translational Systems Science Series* (Vol. 1, pp. 91–128). Tokyo: Springer Japan.

Laszlo, E. (1996). *The Systems View of the World: A Holistic Vision for Our Time*. Cresskill: Hampton Press.

Lessig, L. (1999). *Code and Other Laws of Cyberspace*. New York: Basic Books.

Maner, W. (1978). *Starter Kit in Computer Ethics*. New York: Hyde Park.

Meadows, D. H., Meadows, D. L., Randers, J., & Behrens, W. W., III. (1972). *The Limits to Growth*. New York: Universe Books.

Moor, J. H. (1985). What Is Computer Ethics? *Metaphilosophy, 16*(4), 266–275.

Mosco, V. (2014). *To the Cloud: Big Data in a Turbulent World*. Boulder: Paradigm Publishers.

NASA. (1968). *Earthrise at Christmas*. Picture of 'Earthrise' over the Lunar Horizon Taken by the Apollo 8 Crew on 24 December 1968. www.nasa.gov/multimedia/imagegallery/image_feature_102.html. Accessed [9 October 2017].

Nissenbaum, H. (1998). Values in the Design of Computer Systems. *Computers in Society, 28,* 38–39.

Olivetti. (1965). *Alle origini del personal computer: l'Olivetti Programma 101.* Olivetti: storia di un'impresa, www.storiaolivetti.it and Archivio Storico Olivetti, www.arcoliv.org. Accessed [9 October 2017].

Parker, D. (1968). Rules of Ethics in Information Processing. *Communications of the ACM, 11*(3), 198–201.

Patrignani, N., & Kavathatzopoulos, I. (2013). The Brave New World of Socio-Technical Systems: Cloud Computing. In T. W. Bynum, W. Fleischman, A. Gerdes, G. M. Nielsen & S. Rogerson (Eds.), *The Possibilities of Ethical ICT*, Proceedings of ETHICOMP 2013—International Conference on the Social and Ethical Impacts of Information and Communication Technology, Print & Sign University of Southern Denmark, Kolding.

Patrignani, N., & Whitehouse, D. (2014). Slow Tech: A Quest for Good, Clean and Fair ICT. *Journal of Information, Communication and Ethics, 12*(2), 78–92.

Pruchnic, J. (2013). *Rhetoric and Ethics in the Cybernetic Age: The Transhuman Condition*. New York: Routledge.

RRI. (2015). *Responsible Research and Innovation*. https://ec.europa.eu/programmes/horizon2020/en/h2020-section/responsible-research-innovation. Accessed [9 October 2017].

RRI-ICT. (2015). *Responsible Research and Innovation*. https://ec.europa.eu/digital-agenda/en/news/workshop-rri-ssh-ict-related-parts-horizon-2020-wp16-17. Accessed [9 October 2017].

Science. (2009, July 24). *Complex Systems and Networks*. Special issue, *325*(5939), 357–504.

Shannon, C. E., & Weaver, W. (1949). *The Mathematical Theory of Communication*. Urbana: The University of Illinois Press.

Spiekermann, S. (2015). *Ethical IT Innovation: A Value-Based System Design Approach*. Boca Raton: Auerbach.

Stahl, B. C., Eden, G., Jirotka, M., & Coeckelbergh, M. (2014). From Computer Ethics to Responsible Research and Innovation. *Information & Management, 51*(6), 810–818, Elsevier.

Stilgoe, J., Owen, R., & Macnaghten, P. (2013, November). Developing a Framework for Responsible Innovation. *Research Policy, 42*, 1568–1580, Elsevier.

Systemic Design. (2015). www.systemicdesign.org. Accessed [9 October 2017].

The Disappearing Computer. (1998). Future *and Emerging Technologies*. http://cordis.europa.eu/ist/fet/dc.htm. Accessed [9 October 2017].

TLNF. (2015). *The Long Now Foundation*. http://longnow.org. Accessed [9 October 2017].

Tucker, A. B. (1991). Computing Curricula 1991. *Communications of the ACM, 34*(6), 68–84.

Turing, A. (1937). On Computable Numbers, with an Application to the Entscheidungsproblem. *Proceedings of the London Mathematical Society*, Ser. 2, *42*, 230–265.

Von Bertalanffy, L. (1950). An Outline of General System Theory. *British Journal for the Philosophy of Science, I*(2), 134–165. https://doi.org/10.1093/bjps/I.2.134

Von Neumann, J. (1945). *First Draft of a Report on the EDVAC*. Moore School of Electrical Engineering, University of Pennsylvania, Philadelphia.

Wall Street Journal. (1965, October 15). Desk-Top Size Computer Is Being Sold by Olivetti for First Time in US.

Weaver, W. (1948). Science and Complexity. *American Scientist, 36*, 536–544.

Weizenbaum, J. (1976). *Computer Power and Human Reason: From Judgment to Calculation*. New York/San Francisco: W.H. Freeman.

Wiener, N. (1954). *The Human Use of Human Beings: Cybernetics and Society*. Boston: Houghton Mifflin, 1950; 2nd edn revised, Garden City: Doubleday Anchor, 1954; citations are from the 2nd edn revised, 1954.

Wiener, N. (1961). *Cybernetics: Or Control and Communication in the Animal and the Machine* (2nd ed.). Cambridge, MA: MIT Press, 1948.

Zhou, M. C., Li, H. X., & Weijnen, M. (Eds.). (2015). *Contemporary Issues in Systems Science and Engineering*. Hoboken: John Wiley & Sons. https://doi.org/10.1002/9781119036821

CHAPTER 7

Slow Tech: A Good, Clean, and Fair ICT

Abstract This chapter is the real core of the book. It describes the main characteristics of Slow Tech, a *good*, *clean*, and *fair* ICT—a responsible, sustainable, and ethical approach to ICT. Slow Tech is proposed as a compass for new directions in technology design and use and is inspired by three fields—a reversal of the values implicit in the Olympic motto, the concept of responsible research and innovation, and the Slow Food movement. The importance of each of these fields is explored. First, the framework of the Olympic motto, rooted in the thoughts of Italian activist, Alexander Langer, is challenged. Langer's commitment to acting and working in a way that is *slower*, *deeper*, and *sweeter* is absolutely consistent with a Slow Tech approach. Second, the concept of responsible research and innovation is about products and designs being socially *desirable*, environmentally *sustainable*, and ethically *acceptable* is explored. Third, and central to the volume, the Slow Food movement has its roots in the three principles: *good*, *clean*, and *fair*—each of these three elements is explored in detail and mapped to ICT.

Keywords Acceptable • Balance • Clean • Compass • Deeper • Desirable • Ethics • e-Waste • Fair • Good • Human beings • Learning • Mind • Movement • Reflection • Slower • Sweeter • Slow Tech • Sustainable • Values • War • Workers

© The Author(s) 2018
N. Patrignani, D. Whitehouse, *Slow Tech and ICT*,
https://doi.org/10.1007/978-3-319-68944-9_7

Creating and building a Slow Tech approach provides people with an opportunity to work together on who they are and who they want to be both in terms of today's world and the world to come for future generations.

This period of reflection is a particularly important step. This is because, in terms of human beings' relationships with technology, society is now on the threshold of a variety of developments. The environment is not yet populated with artificial agents or by humanoids, people's bodies have not been filled with nano-bots and implants (Patrignani 2009a), and it is before society is completely dependent on an ICT infrastructure that is totally unsustainable in the long term (Patrignani et al. 2011): people still have the opportunity to change direction. The option still exists to think and act consciously about the challenges that are entailed in developing and designing the next generation of technologies.

People can still grasp the chance to learn how to use ICT wisely and to begin to prepare the foundations for a new sense of digital wisdom (e.g. Bradley 2017). In the context of this book on Slow Tech, digital wisdom is taken to mean a human approach to the design and use of ICT that takes into account human beings with their limits, values, and desires as well as the dimension of time. These considerations enable a reflection on everything that is related to the responsibility, sustainability, and ethics of an approach to ICT.

7.1   Introducing a Slow Tech Compass

To find a route forward, a compass is needed. The compass implies that there need to be some basic principles outlined that could help people in their search for a new technological future.

In order to introduce a Slow Tech compass for finding a route forward in ICT, certain assumptions need to be challenged. These considerations tend to be ones that are often taken for granted in contemporary society. There are various examples of such expectations: they include an improvement in the speed at which technologies are used regardless or an enhancement in productivity or quality of life is achieved either by ever faster ICT or by the provision of more information.

In contrast, it is important to show that there are alternatives to this preoccupation with speed and with information overload. Instead, it really is possible to identify actionable actions towards Slow Tech. As a result, positive Slow Tech guidelines or evaluation tools could be formulated.

These guidelines could help people generally to make sound decisions. They could be especially useful for people when they are faced with a new technology, a challenging new scenario, or a context which is dramatically changed by the introduction of a new disruptive technological innovation. They could also be used by designers and engineers when they start to develop a new system based on information technology.

This set of options and alternatives makes it obvious that there should be widespread contemporary public debate about such emerging technologies as big data, the Internet of Things, and implants/close-to-body technologies. It is also evident that these kinds of reflections will enable innovative, alternative propositions to be put to various stakeholders, such as the ICT industry, computing professionals, communities of users, and policy makers. These alternatives could include a collection of proposals and recommendations for a more human-centred and more future-proof ICT.

7.2   ADAPTING THE PRINCIPLES OF THE SLOW FOOD MOVEMENT TO ICT

What can be learnt from the Slow Food movement that can be introduced into new ICT scenarios? When people use an ICT system or product or service, on what issues should they reflect?

The Slow Food movement originally introduced a reflection on the whole food chain. Similarly, we propose a Slow Tech approach that begins a reflection on the entire ICT value chain. This means adopting the *good, clean,* and *fair* approach to food proposed by the Slow Food movement and applying it to ICT.

A number of parallels can be seen between the two sets of thinking. The triplet of ICT needs, with its emphasis on *good, clean,* and *fair,* acts as an invitation to search for a new balance between rational thinking (what is *clean* and *fair*) and democracy (what is *good* in an equitable sense), in which dreams, arts, and beauty (what is *good* in a pleasant or aesthetic sense) play a fundamental role.

These modifications to the use of ICT—Slow Tech—may permit a return to a more leisurely pace, in which it would be possible to work with pleasure and a sense of life and where ICT could support individuals and society through a wide variety of ways to work and act collectively and collaboratively. The network experience would become a new space in which people could brainstorm, develop ideas together, and participate in democratic initiatives both online and in real life. This concept of Slow Tech could help people to expand their focus onto the whole of human life cycle, from birth to death, so that they do not simply try to fit a part of their lives, based on work, into a 35-year time period. As a result, ICT will become a companion through the entirety of human lives.

In the subsections of this chapter, there is a focus on the three elements of good, clean, and fair ICT. Good ICT is dealt with in the greatest detail. Ultimately, both clean and fair ICT deserve equivalent treatment. Some of this effort has been made in other publications: for example, clean ICT is dealt with in a special issue of an ethically related journal, the *Journal of Information, Communication and Ethics* (Patrignani and Whitehouse 2015).

7.3   GOOD ICT

Good ICT means an ICT that is human-centred; when designers place human beings centre stage and start designing ICT systems for them, they realise the importance of taking human limits into account. Such in-depth investigations will enable people to consider more carefully the entire range of implications of the introduction of technology. As a result, they will design more complete stakeholders' networks that involve the whole span of people who are prepared to face the challenging dilemmas related to ICT.

So a good ICT is a *socially desirable* and hospitable ICT. It can support people in balancing working time and leisure time by facilitating a slow lifestyle, that will also facilitate learning and education, and will be respectful of the nature of human minds.

7.3.1   *Good ICT as Socially Desirable ICT*

In recent times, there has been a tendency to concentrate just on the economic and rational sides of ICT. In today's ICT applications, and even in society, many of the desirable aspects of technology have been lost with the exception of a certain number of design considerations. An understanding of the full completeness of life has been disregarded.

Identifying what is socially desirable means capturing not only what the individual desires but also what is desirable for other people. In order to find out what other people desire, discussions and dialogues need to be held in a serious attempt to reach consensus and common understanding.

The process of undertaking conversations together will help to develop what is desirable as a whole. The good news is that there are technologies to facilitate this: blogs, wikis, and social media.

Open innovation and design is beginning in many fields of society that range from industry (Von Hippel 2006; Sigismund Huff et al. 2015; Chesbrough 2015) to policy makers and governments (Van Abel et al. 2011).

An innovation will tend to be socially desirable if it is based on open innovation and driven by a community of shared interests built up between designers and users. Essentially, this kind of commonly shared process enables identification of what is desirable to large numbers of people.

Many companies and organisations are starting up such communities among their clients and users. From the information shared among the discussants in these communities, they define the main possible options for new directions in terms of their research, products, and services. The role of networks in these open innovation communities is fundamental: people gain access to information by using huge knowledge repositories, they exchange ideas in social networks, and they propose innovative ideas to the organisations involved. In the automotive world, for example, users can contribute to the design of new cars. One example of such a contributory initiative ran every two years between 2000 and 2007. Peugeot Concours Design was introduced in 2000 and encouraged users to submit their design for a car. The winning design was unveiled at the biennial Frankfurt motor show (Peugeot 2015).

The first requirement when developing socially desirable ICT is the ability to collect a set of proposals and desires from society at large. Processes and tools are, of course, needed to facilitate this approach. This implies the creation of a dialogue, that is, a fruitful, new interaction between the various innovation actors, such as the researchers, designers, engineers, and social actors involved in the creation of any technology. It is very important to include end users in the design process from the beginning, including during the stage of idea generation. This will create unavoidable conflicts (Habermas 1996).

With many-to-many communications and unlimited access to knowledge, designers could improve citizens' participation in policy development; ensure that people can take deeper, broader, and more informed decisions; encourage transparency by governments; and include disabled and elderly people in all forms of decision-making so that society heads towards greater real e-democracy.

How can what is socially desirable be defined therefore? A number of conditions might be considered under the umbrella term of socially desirable technology. As an example, ICT should be designed in a way that is accessible to all in order to address a wide spectrum of people's needs and, in particular, their special needs. Here, e-accessibility, e-inclusion, and Design4all approaches are all appropriate. Good ICT could prove useful for designing highly personalised, multimodal interfaces for people with special needs. A multimodal interface provides end users with ways of interacting with a computer system, through several alternative channels and tools to use for both input and output data. This is a basic concept for overcoming sensorial barriers. It requires having a very sophisticated profile of the person who sat or stood in front of the machine or encouraging the person to use the technology closer to their body.

Good ICT is when ICT is adapted to human beings, and not vice versa. This means respecting people's privacy, dignity, and autonomy. One example illustrates *sweeter* ICT: in an ambient intelligent environment that supports elderly people in their own homes, good ICT could mean providing individuals with the capability to switch off a monitoring system since, while these persons have the right to be helped, they also have the right to be left alone. An example is located in the Regione Friuli Venezia in Italy (a region with 25 per cent of population over 65 years old, while the Italian average is 21 per cent) with its initiative on Active and Healthy Ageing in Information Society (FVG 2016).

Good ICT could also provide a medium for social and cultural integration by enabling minorities, migrants, and marginalised young people to be integrated fully into communities and participate in society. It could act as a fantastic opportunity for geographic e-inclusion, increasing the social and economic well-being of people in rural, remote, and economically disadvantaged areas. It could offer a way to empower older people to participate more fully in the economy and society, continue to have independent lifestyles, and enhance their quality of life and indeed their well-living. In general, digital competences will become more and more important in order that citizens are equipped with the knowledge, skills, and lifelong learning approaches needed to increase social inclusion, employability, and the enrichment of human lives.

Hence, it is important to understand that the costs and benefits of any technology may today be distributed in quite an imbalanced way among all the players involved in society. The parties which are involved in making initial investments may not necessarily be those who derive the greatest benefits societally or economically. If organisations aim for too rapid return on investment, they may be disappointed by the results and may be deterred from making similar types of investments in the future. A positive move could, instead, involve consideration of what is known as social return on investment (Cabinet Office 2009).

So too, there could be a regeneration of concern for positive human behaviours. In today's consumer society, many people are accustomed to having items and applications available to them whenever they desire, often at periods of peak demand. Indeed, the strength of contemporary marketing messages sometimes induces unnecessary needs that are generated top-down and reflect the technology push of industry. People continue to want ICT goods and services that are faster, more powerful, and ever more available. In their search for short-term gratification, they often fail to see the long-term implications of their actions.

People need, however, to make conscious choices about exploring what is best about human beings in terms of their qualities, such as work and life balance, care, cooperation and collaboration, kindness, mutuality, reciprocity, and sharing in the longer term, rather than simply aiming for that which is instantaneously gratifying in the short term.

In the early part of this decade, some interesting approaches were already explored when public sector groups tried to design socially desirable future ICT scenarios, using participatory approaches. See, for example, the Futurium Project of the European Commission, to which people could contribute freely to build a vision for a 02050 ICT scenario (European Commission 2012b).

7.3.2   Good ICT as Hospitable ICT

The concept of hospitable computer systems is crucial to good ICT. This term is borrowed from the Japanese expression of *omotenashi* (Kanamaru et al. 2015). Hospitable ICT is an ICT-related environment that is not just cosy but that is also human-aware or human-centric.

By focusing on hospitality, it is possible to recognise that designing ICT environments around human beings implies slowing down the entire design process. Indeed, in some ways, the technology itself will be slower or at least its speed will no longer be its most important feature. Indeed, as futures thinker Jonathan Margolis implied, it is otherwise easily possible to reach a state of terminal velocity (Margolis 2001).

Hence, it is feasible to introduce the concept of slower technology as a fundamentally social issue: why not begin to take human beings as the starting point, the central point for the design of technology? Technology should not be created to improve human performance but to enhance the well-being and well-living of persons and communities. This approach would introduce a shift whereby people will no longer concentrate on the clock speed of technology. Instead, their focus will be on the flow of processes involving human beings. As a consequence, there would no longer be a need to change computers or phones every 18 months or fewer. Instead, people would need to spend more time on the actual design of ICT systems. A participatory design approach to ICT will be needed that brings end users into the design team at the very beginning of a new initiative or project. For example, there is now evidence of interesting experiences in which young people teach older adults how to use computers and software applications and how to go online called 'adopt an elderly person' (ISF 2015). This approach demonstrates that, when a younger teacher slows down any explanations and takes more time to use interesting metaphors, the older learner appreciates these intergenerational exchanges of knowledge and learns much more easily (Spector et al. 2014). It has also been shown that, when people have enjoyable and stimulating experiences, they live longer (Crawford 2004). Since the older adult may be able to move using only slow movements, it is important for a teacher to talk slowly and to use slower communication protocols (ISF 2015). As a result, the younger person/teacher discovers the importance of adapting a technology environment to the older human beings with whom he or she is collaborating.

7.3.3   Good ICT and a Balance Between Leisure and Working Time

ICT can also be good when it helps people to find an appropriate balance between their work time and their free time or leisure.

In Latin, such free time was referred to as *otium* (leisure or idleness); *negotium* was the contrary. Thus, *otium* meant an activity dedicated to intellectual speculation, time for a rest or free time that was given over to private life and studies. *Otium* could be contrasted with *neg-otium*, that is, exactly its opposite, meaning not-leisure (and which, in Italian, is the origin of word *negozio* that means a shop or store). Hence, *negotium* is an activity dedicated to business activities or to public affairs (Cicero 1998).

Over time, on the one hand, *otium* has gained a bad reputation and now incorporates a negative meaning related to laziness or sloth. On the other hand, *negotium* has now absorbed many people's lives almost completely while leaving others under- or unemployed or employed in work that has little meaning.

Leisure time and idleness is the time needed by people for themselves, and people's minds and bodies use to recuperate. Leisure time is the time needed for intimate and personal needs ('being'), to improve human well-being. It is the time in which to regain a sense of balance. An important aspect of leisure is that it is voluntary and is based on free choice. If the leisure time is not selected through free choice, then it could be equated with, on the one hand, unchosen unemployment or, on the other hand, could become a compulsory, commercial activity (e.g. many commercial holiday packages are limited to specific time periods and are pre-organised in all their details).

Today, these are difficult times. The notion of free time is changing. Nowadays, the risk is that ICT is accelerating people's daily lives and transforming their being (their pleasure and their leisure time) simply into repeatedly doing. Business has developed to increase people's possessing goods and financial assets. ICT has the power to manage immense quantities of information, yet human beings' time and attention are not considered to be scarce resources (Goldhaber 1997). In organisations, the design of ICT, software applications, and human-computer interactions continues to concentrate just on managers' short-term commercial results and gains; they rarely take into account any limiting factors (Davenport and Beck 2001). The always-on capability enabled by ICT runs the risk that ICT will have major impacts on people's lives: for example, a 2012 research study on this subject showed that only a minority of organisations has a formal work and life balance policy in place. Only one per cent of the organisations studied had 'days or time when email is not used' (e.g. mail-free Fridays or mail-free weekends) (SHRM 2012).

ICT should in fact play a much more constructive and positive role in terms of finding a balance in people's lives, rather than simply focusing on the business and commercial aspects of society. Instead, ICT should help people to regain a good equilibrium between the two dimensions of their lives, that is, leisure and business or life and work. This new kind of ICT would be *good* ICT: human-centred ICT, technologies that improve human well-being and well-living (well-living is a play on words invented by the authors, intended to focus on the idea of living well). It would be a wise move to find the appropriate balance between *otium* and *negotium*, between the time needed for ourselves and the time required for work, family duties, and obligations (Covey 1989, 1996).

When Slow Food defines a food as good, it implies that the food is also delicious and pleasant (Petrini 2011). It refers to the dimension of pleasure: not only is the sense of taste stimulated, but there is also a much more complex and enjoyable involvement of all the senses and the entire body and mind. Can one imagine a similar experience with ICT? How can human interactions with computers be transformed into more enjoyable experiences?

ICT systems that are good for human beings might be considered to be good when the systems are designed by using a human-centred approach. In the case of human-centred, good means good for human beings, that is, good for their being. Thus, ICT systems would incorporate qualities that are strongly related to and include human histories, memories, and values. If good means helping people to enjoy all the senses of their bodies, then ICT, of course, would also be good when it supports people to handle certain diseases or when it specifically helps elderly people and people with disabilities in the fields of e-health and e-inclusion.

Any process that involves human beings and computers will be the result of a complex interaction. This complex interaction can be an enjoyable experience only if the system and the human-computer interfaces are designed by taking into account human limits.

7.3.4   Good ICT and Slower Life

Many thinkers, poets, and artists try to introduce a longer-term view of life. One noteworthy example is the work of the musician, John Cage (1912–1992). His composition for the organ, *As slow as possible*, is performed by just a single solemn note that is played every year. The piece is currently performed at the organ of St. Burchardi church in Halberstadt, Germany. It was first played in 02001 and is scheduled to last for 639 years, ending in 02640.

Faster appears to be the mantra that surrounds much of the ICT development, use, and outcomes. Each new generation of electronic gadget promises to work faster than the previous one. The characteristic of fast can be attractive to people as consumers not just because of their fascination with speed but also because it is usually combined with even more technical features that produce more powerful and, typically, cheaper products (Lennefors 2013).

People can stay always on, becoming just a node of a network. However, do they really always need to be connected? Do they need to stay online for 24 hours? Do they have to have all their input channels permanently stimulated? Rather, human beings need time and silence for the purposes of thinking, creativity, and imagination.

Human beings are now realising that the speed of their bodies and minds is not compatible with the speed of many computer-based systems. Hence, it is important for them to challenge the mantra of speed and to modify it. On the one hand, human beings need time and silence to think. On the other hand, they need time to concentrate on more specific human aspects when designing ICT systems. They also need to bear sustainability in mind—taking into account the limits of Moore's law (Moore 1965; Patrignani et al. 2011).

In some organisations, the mantra of speed is indeed starting to be questioned. Managers and employees are starting to question the 'clock' around which company processes are based since, if these processes are increased to the extreme, they can lead to inappropriate processes and results, wasted time, and unhappy workers (Burton 2013). These observations are particularly valid in the case of knowledge workers (McGinn 2011). Knowledge-based enterprises are aware that their main assets are their employees, who can be called knowledge workers. In order to retain such knowledge workers, companies need to provide them with an ICT environment that facilitates knowledge enrichment and knowledge exchange; companies cannot just squeeze their employees by keeping them constantly in knowledge delivery mode. Hence, organisations are introducing more enjoyable (and more productive) time management, 'results-only work environment', self-controlling time and calendars, office presence no longer required, with an emphasis on more holistic results (Ressler and Thompson 2008).

Important organisational aspects such as productivity, long-term sustainability, and the quality of the processes at the core of the organisations seem no longer to be directly connected just to computer speed. At the core of all organisations' processes today is one basic infrastructure: computers and networks, hardware and software.

Although technical developments in society are moving very fast, there are benefits to slowing down and using decision-making processes that involve and incorporate larger groups of people. Getting people to describe their own ideas enables reflection, facilitates the absorption of the ideas of others, permits the capacity to address ethical dilemmas and to identify them, and allows people who think perhaps more slowly or who absorb ideas more gradually to get on board an initiative. Important insights into how differently people perceive the world and make decisions that can help were made by Isabel Briggs Myers and her mother (Briggs Myers 1987).

One example of a slower form of decision-making is the Participatory Design approach, where the design team includes end users from the beginning of the design process. In software design, the Computer Professionals for Social Responsibility (CPSR) was a pioneer in this field: 'Participatory Design (PD) is an approach to the assessment, design, and development of technological and organizational systems that places a premium on the active involvement of workplace practitioners (usually potential or current users of the system) in design and decision-making processes' (CPSR 2008). The movement was inspired by Scandinavian experiences that took place in the 01990s (Bjerknes and Bratteteig 1995; Bødker 1996).

This form of design method, based on participation, probably implicitly involves a shift away from representative democracy to a more shared form of decision-making. The two types of decision-making can be called autonomous and heteronomous. Today's pervasive heteronomy approach is much faster than an approach based on autonomy. Heteronomy implies rule by another, which means that one person delegates to another person critical decisions such as ethical choices. In contrast, autonomy means self-government and freedom from external restraint (Kavathatzopoulos 2011). Of course, developing autonomy needs time. Autonomy is slower than heteronomy because it takes time to reflect. To be respectful of the autonomy of people, it is better to slow processes down. As a motto, one could say 'To improve autonomy, slow down'.

7.3.5   Good ICT and Learning

Good ICT could also mean that people are able to access more interesting sources of information so that they become more intelligent. Thus, they are more able to make connections and cross-check sources of information. Today, ICT enables people to scan large quantities of information, but in a very superficial way. While examining the immense possibilities and consequences of browsing the Web quickly, people run the risk of being washed away by a waterfall of bits. The flood of information that surrounds them does not necessarily enable people to take the time to really understand the subject matter, compare the data with their own internal codes, and develop a sense of critical thinking.

Browsing the Web provides a huge opportunity for people to improve their access to knowledge, but—by itself—this is not enough. This is why it is important for teachers to suggest to students to provide not just a Webography but also a bibliography when writing their reports and essays. A 02015 study by the Organisation for Economic Co-operation and Development (OECD) demonstrates that at school teenagers who are strong in reading, which is typically a slow process, are also good at developing critical thinking, obtain high results in all subjects, and can select in an effective way links from search engines so they are sound Web navigators (OECD 2015). When searching the Web, it is easy to observe thousands of links among documents. Yet, there are few sources of informed, intelligent help that can facilitate the selection of results based on people's own ranking or on rankings that are founded on scientific and transparent algorithms (rather than on rankings biased by the commercial titans of the Web search industry) (Gotterbarn 2015). There is a current and growing preoccupation with the difficulties posed by this mass of big data (Floridi 2012). Ultimately, there is a risk that people become digital consumers who superficially navigate clouds on the surface of touchscreens rather than maturing into the digital citizens that ICT could instead promise them to become.

Digital natives, and millennials, are good at using ICT. Yet in many schools, teachers are discovering that millennial students have great difficulties either in writing or in pronouncing longer or more complex sentences (Palfrey and Gasser 2008). Digital natives use short, simple sentences, without benefitting from the ability to coordinate phrases or join up phrases using subordinate conjunctions. The *parataxis* style of communication, where only simple and concise sentences are used, is beginning to dominate (De Carli 1997). Phrases are listed without any coordination or subordination of connectives. All the phrases are presented at the same level. In contrast, the *hypotaxis* style of communication is becoming increasingly unusual: sentences are no longer as rich or complex, contain subordinate expressions, or include phrases which are not all at the same level. This second approach is a deeper way of writing and speaking that can be cultivated, for example, by reading good books (or e-books) rather than just by tweeting or blogging. In contrast to undertaking broader or more cumulative word and knowledge searches, when writing or speaking, today it is—in contrast—important to aim to go deeper.

7.3.6   Good ICT and Studies of the Human Mind

There are many dilemmas surrounding the electronic communications that are taking place inside and outside tertiary care and other institutions. The dilemmas apply to people of all ages, not simply older adults. For example, in the area of brain imaging or neuro-computing, researchers are working on developing and understanding data that can be gathered from various imaging techniques, such as positron emission tomography (PET) and functional magnetic resonance imaging (fMRI). It can permit the analysis of computer-generated brain images (Filler 2009). This research area carries many new promises, but it also poses great questions and ethical dilemmas since it enables unique access to views of both the conscious and unconscious processes of the human brain. On the one hand, this kind of research is sometimes undertaken inside research laboratories that are dedicated to medical applications and are subject to strict protocols and regulations developed by ethical committees. Their results promise to throw light on brain disease through studies in pharmacology, lesion localisation, and functional association. Research in these fields is subject to strict ethical review and approval even if many of the research questions are still fundamentally open. On the other hand, there are research initiatives that are undertaken outside of medical laboratories that are pursued purely for commercial application. Here, perhaps even more ethical dilemmas arise. The use of applications such as lie detectors, brain fingerprinting, neuro-marketing, and screening tools for job applicants and school leavers is opening up a long list of new, ethical questions (Patrignani 2009b).

Many questions arise. Can the human mind be studied as a standalone (brain) machine? Can the brain be disconnected from the mind-environment information flow (Bateson 1979)? Can the image of an intention be captured? Are computer-generated images really needed to diagnose psychiatric illnesses?

Answers are beginning to be untangled. For example, intention does not imply behaviour or action. The mental exploration of a desire may reflect a form of play acting rather than a will to explore the desire in practice. Some psychiatric diseases and conditions cannot be treated without special care and without dedicated communications and human interactions. Human beings who have psychological and psychiatric conditions need warm support and rich interactions with carers, social assistants, and friends and family. Perhaps ICT will bring useful information to doctors, and clinical and medical support staff, so as to improve medical treatments.

However, people cannot be reduced simply to such notions as their brains, their nervous cells, and interconnections. The human mind is not a computer based on wet-ware, it is the emerging property of the complex mind-environment information flow.

7.4   Clean ICT

ICT is *clean* when its systems are designed and produced by taking into account the impact of technology on the planet.

In times past, ICT was considered to be an island: it was a technology that was always accepted since it was perceived as being new and clean by its very definition. Today, rather, it is proposed to see ICT according to a bigger picture: by taking into account the entire life cycle, from raw materials, to ICT use and applications, to e-waste management and recycling (Patrignani et al. 2011).

An innovation is assumed to be an idea that becomes a real product and service that is manufactured and distributed by actual companies to genuine people. An innovation will be *environmentally sustainable* if the design, manufacturing, and recycling processes for introducing products and services into society take into account the innovation's impact on the planet. Nowadays the impact of the entire life cycle of a particular product or service should be carefully examined in advance of its production. Such an approach is becoming mandatory for companies with long-term strategies. In particular, it is evident that not only are environmental issues on the management agendas of companies but sustainability initiatives are too. These approaches can in fact be profitable: simply speaking, companies' customers are beginning to prefer sustainable products and services (MIT 2012).

This way of thinking about the long-term impact of actions on the environment, originally introduced over 50 years ago, is becoming increasingly diffused throughout society. Human beings are starting to think that innovation must be not only socially desirable but also *environmentally sustainable*. For many thinkers, well-being cannot just be the immediate satisfaction of desires; it cannot be disconnected from the well-being of the planet and future generations. Rather, responsible innovation requires that ICT must be environmentally sustainable.

How can people enjoy their well-being today if they are conscious that they are compromising the well-being of the Earth itself and future people? How can ICT be designed to take into account the environmental limits of the planet? What does environmental sustainability mean for ICT? Is it possible to develop sustainable ICT? Surely the long-term sustainability of ICT needs to address issues such as the production, manufacturing, powering, and recycling of ICT. In terms of producing ICT: where do the rare-earths necessary for chip manufacturing come from? For how long can these reserves be relied on? In powering ICT: if the energy necessary to power the gigantic data centres of the cloud computing era doubles every five years, how can this growing amount of energy be produced? For how long can the related growth of CO_2 and subsequent climate change be tolerated? Many studies have investigated the challenges of the growing demands for ICT infrastructure (Fettweis and Zimmermann 2008) or researched zero-power ICT (European Commission 2012c).

What happens when dealing with the waste and recycling of ICT products, that is, recyclability-by-design? For example, the open hardware movement looks very promising for the future (Lahart 2009). It will help the invention of new ways of designing hardware based on the recyclable-by-design concept. The open hardware approach enables the collection of thousands of contributions from experts that will facilitate the challenges of sustainable ICT to be addressed.

7.4.1   Clean ICT and Climate Change

Several international studies over the past several years have investigated the enabling of the low-carbon economy in the information age. One of the most recent (GeSI 2015) shows that, even if ICT itself contributes to CO_2 emissions with 1.25 $GtCO_2$ (0.59 due to end user devices, 0.3 to voice and data networks, and 0.36 to data centres), by 2030 it could contribute to CO_2 reduction by 12.1 $GtCO_2$ (1.8 in power generation and distribution, 3.6 in transportation and mobility, 2.0 in agriculture, 2.0 in buildings, and 2.7 in manufacturing). So it looks like the balance is a significant reduction of 10.85 $GtCO_2$. Indeed, further research is needed to take into account the impact of CO_2 due to the production of ICT devices and the growing mountain of e-waste.

However, even if the challenges of the three-stage life cycle of producing-powering-recycling ICT can be dealt with, can society afford the speed of this cycle? Is it feasible to continue to want to exchange electronic gadgets every 12 months (Patrignani et al. 2011)?

Again, time and speed reside at the core of these reflections about ICT. This emerging need to slow down ICT consumption cycles lies at the heart of the Slow Tech concept.

7.4.2 ˙ Clean ICT and e-Waste

Clean ICT should also consider seriously the destination of hardware at the end of its use. Despite the publication of the European Waste Electronic and Electrical Equipment (WEEE) Directives on e-waste (European Commission 2002a) and the Restriction of Hazardous Substances in electronics (European Commission 2002b) 15 years ago, at the global scale electronic waste remains one of the biggest ecological problems still to be tackled. At an international level, the majority of e-waste goes to unknown destinations. Its precise treatment is unclear and there is a high risk of environmental pollution due to the hazardous substances treated, like lead, cadmium, chromium, and mercury. *The Wall Street Journal* journalist Cris Prystay termed e-waste 'the world's fastest growing and potentially most dangerous waste problem' (Prystay 2004).

For all these kind of reasons, Greenpeace, one of the world's most recognised environmental advocacy organisations, has, since 2006, been monitoring the ICT industry strictly. Now in its 18th edition, Greenpeace produces a *Guide to greener electronics* (Greenpeace 2012) by ranking three actions demanded of electronic companies: clean up their products by eliminating hazardous substances, take back and recycle their products responsibly once they become obsolete, and reduce the climate-related impacts of their operations and products. Each score is based solely on public information available on the electronic companies' websites. Even the cleanest ICT industry and the most efficient recycling mechanism cannot, however, cope with the growing speed of ICT consumption. Greenpeace's last recommendation about ICT is therefore: 'Remember! The most sustainable devices are the ones you don't actually buy! Work to extend the life of your existing electronic gadgets, buy used products, and only purchase what you truly need' (Greenpeace 2012: 1).

7.5   Fair ICT

How is it possible to define *fair* ICT? The ethical reflections around ICT can be considered a fundamental component of a *fair* ICT. Fair can mean right or just, or it can imply taking into account the interests of all the stakeholders involved. These aspects of fairness—the ability to add to the sense of equity or justice or the capacity to increase the socio-economic benefits to society—are explored in this subsection. While it would be feasible to explore fair ICT in much greater detail, this part of the chapter explores just few domains related to fairness: at one level, ethics and shared values; at a more applied level, the rights of workers and the health of workers.

7.5.1   Fair ICT and Ethics

Fair ICT needs to be ethical ICT or at least ethically acceptable: fairness as a characteristic shares much in common with the notion of responsible innovation. It is dependent on an understanding of ethics.

Working in this field implies the need to focus on computer ethics, and using specific tools and techniques can address either new ethical dilemmas or old ethical dilemmas in new contexts (Johnson 2009). Providing a list of the potential ethical dilemmas facing humankind would be a long and substantial process. Even creating a short list is challenging.

It is, nevertheless, useful to outline a few illustrative questions that arise when developing such a list. Here are almost 20 examples: What should be considered to be right or wrong when designing an e-democracy application? What is right and what is wrong when designing systems for people with different abilities, cultures, and languages? What issues arise when introducing computerised processes into workplaces? Or when introducing ICT into an educational environment? Who will select what is worthwhile content for pupils or students to study? Can that choice be delegated to search engines? How will a four-century-old copyright system evolve in a situation in which ICT is introducing immense changes and where—for example—the copy is itself the original? How should the intellectual property right system be changed in such a way that innovation is stimulated and digital natives or millennials can express their creativity freely? Is the creative commons the most appropriate approach? (Creative Commons 2015). How can the security and reliability of ICT systems be improved by leveraging the skills and competences of people who are really experts in this field, like hackers? How can privacy-by-design concepts be introduced in the development of ICT in order to guarantee the control of sensitive data (i.e. the *habeas data* principle)? (Rodotà 2005). How can computer crimes be prevented when these crimes are victimless and when the majority of people who are guilty actually work inside the organisations that are the targets of the crimes? How can society become responsibly informed about the dirty secret of software unreliability? How and when does it make sense to introduce implants or nanotechnologies inside human bodies? When is it appropriate to use artificial intelligence applications? Do human beings really want to apply ICT systems in war scenarios or to pursue research to construct robot warriors?

From the design aspect, fair ICT needs to have been developed through the use of an open and more inclusive ethical governance model that will enable the handling of ethical dilemmas generated by the development of complex ICT systems (EGAIS 2012).

This range provided of 20 very different dilemmas—that cover fields as diverse as democracy, education, copyright, implants, and weaponry—requires a set of specific steps. These steps include an in-depth analysis of the context, the construction of a complex stakeholders' network, the identification of all the relevant social and ethical issues in order to examine all the possible alternative scenarios, and the development of appropriate advice for policy makers, designers, organisations, and end users. In the case of any profound ethical dilemma surrounding ICT, the analysis needs to include all the issues raised by all the actors and stakeholders, complete with their varied histories, cultures, values, and desires (EGAIS 2012).

There is no magic formula for solving the ethical dilemmas that can emerge from a technological development. It is, however, possible to propose an inclusive process that starts from shared values rather than from abstract theoretical or ethical principles.

It is very important to underline this lack of a precise formula, because it has implications for the need for a set of competences, skills, capacities, and processes needed to face ethical challenges. For example, if engineers are designing an ambient intelligence system that is due to take care of elderly people in their homes, is it right or wrong to include a 'power off' command that can be used by people who would, in some situations, like to be left alone? Can the engineers really address this design issue without involving the real people for whom they are designing the system? For an active illustration of such points, see, for example, Clara Berridge's exploration of a long-established home monitoring system in the United States of America (Berridge 2014).

A European Commission co-financed project, called ETICA (ETICA 2011), addressed the main ethical issues related to ICT that are likely to emerge over the next decade. They are associated with a number of technologies that include ambient intelligence, augmented/virtual reality, future

Internet, robotics, artificial intelligence, affective computing, neuroelectronics, bioelectronics, human-machine symbiosis, cloud computing, and quantum computing. In order for society to be prepared to address any ethical dilemmas connected with these technologies, the ETICA project provided a collection of recommendations for policy makers, industry, and researchers. The suggestions for policy makers included the provision of a regulatory framework which would support Ethical Impact Assessment for ICT and the establishment of both an ICT Ethics Observatory and a forum for stakeholder involvement. The suggestions for industry and researchers included the incorporation of ethics into ICT research and development and the facilitation of ethical reflexivity in both ICT projects and ICT practice. In terms of research activity that is co-funded by the European Commission, all ICT research activities that are intended for financial support are subject to potential ethical reviews: 'Ethics Review is an important part of the process undertaken by the European Commission when evaluating research proposals ... All applications that have been pre-selected ... and that raise ethical issues must be submitted to an ethics review' (European Commission 2012d).

An important contribution to the domain of ethical ICT has been provided in some countries by the introduction of codes of ethics for computer professionals (CEPIS 2015; ACM 2015). Introducing an ICT code of conduct—a voluntary set of rules which people would agree to follow or abide by—is a process that has the benefit of opening up in-depth discussions among ICT professionals about their responsibilities when dealing with their customers and society (Gotterbarn 1992; Whitehouse et al. 2015). One of the most interesting examples of such an approach, from an ICT perspective, is the Software Development Impact Statement (SoDIS) methodology. This methodology provides a responsible ICT project risk analysis. It is useful for identifying the potentially negative impacts of an ICT project and suggesting actions to prevent those negative impacts from occurring (Rogerson and Gotterbarn 1998).

7.5.2   Being Ethical: Working with Shared Values

In its long-term temporal dimension, fair ICT implies showing future ethics (Jonas 1985). Longer-term time dimensions complement the *ethical* dimension.

An innovation will be *ethical* if it is in harmony with the shared values of society. In European terms, in a society aiming to be based on shared values, in order to respond to societal challenges, research as innovation 'must respect fundamental rights and the highest ethical standards' (European Commission 2012a). Ideally, therefore, ethics should be embedded in any research activity and in the generation of processes and activities underpinning knowledge creation.

One of the main challenges with being ethical, of course, is the dependency on the definition of what shared values are. At least one method was proposed by a European Seventh Framework Programme project called EGAIS. The project consortium searched for an appropriate way to identify and address the ethical issues needed for a responsible form of innovation. It investigated precisely the challenge of what is the optimum approach to the ethical governance of emerging technologies and innovations.

The EGAIS project proposed a blending of approaches: procedural (rule-based), reflective (context-based), and substantive (value-based); this blending of approaches was defined as comprehensive proceduralism (EGAIS 2012). It can be summarised in the following five steps:

1. Concentrate on process, not on outcome—create the conditions for a true dialogue between the two fields of science and technology on the one side and society on the other.
2. Involve all stakeholders—find the right merger of persons to include in the dialogue, civil society or end users, and in the research team itself.
3. Expose stakeholders to different scenarios—define a scenario where the proposed activity (e.g. a research project or an emerging technology development) plays a fundamental role in a given context. This may imply also having to undertake a foresight exercise.
4. Co-construct the stakeholders' network—define a network where, in the specified context, the values that underpin the beliefs of the actors, and their complex network of relationships, are likely to emerge.
5. Reflect together with all stakeholders on the different views. Ask questions like 'Why is this value important to me? Why should it be important to anyone?' This final stage of the reflection process will facilitate the movement from individual values to shared values and norms so as to enable the agreement on the meaning of being ethical.

This five-step process is likely to bring about a set of co-constructed norms. As a result, all the actors involved in the particular initiative will be more willing to accept these norms than if they had not undertaken a collaborative exploration of values (EGAIS 2012). It is on this path, first suggested by EGAIS's researchers (EGAIS 2012), that one can hope to find the appropriate way to identify and address the ethical issues that are important and at the heart of responsible innovation.

7.5.3   Fair ICT and the Rights of Workers

Fair ICT means taking into account the conditions of the people who produce ICT, particularly the rights of workers involved in the entire ICT value chain. ICT companies will need to check on business partners' and providers' policies about the working conditions of all the employees or sub-contractors.

Can the ICT industry accept being dependent on certain types of hardware and software providers? What follows are two examples of the unfairness of much ICT production, at two different stages of the production life cycle. The first focuses on ICT industry employees' working conditions, the second explores the mining industry.

Starting from the level of workers in the industry: fair ICT could mean technologies and systems that result from processes that take into account the needs of the people who produce the technologies and the rights of workers involved throughout the entire ICT value chain (SJMN 1985). In 02012, the conditions of workers in the ICT industry reached the headlines of the *New York Times*: many of the newspaper's readers realised for the first time in their lives that their ICT gadgets were produced in factories where men and women were working for only a few dollars a day, for shifts of up to 12 hours a day, for six or seven days a week, and where these employees were allowed only a minimum of rest in dormitories built inside computer plants. In reaction to these horrific working conditions many of the workers began to protest; even more dramatically, some committed suicide (Barboza and Bradsher 2012; Duhigg and Barboza 2012; Duhigg and Bradsher 2012). It was later acknowledged that small improvements may have taken place in Chinese manufacturing companies, although it was still obvious that profound positive organisational changes might require many more decades to occur (Bradshaw and Duhigg 2012).

The tragedy that lies behind one of the raw materials that is most needed in the ICT industry has also received attention: it focuses on the mining industry and its working conditions. Coltan is a mixed material made up of Columbium or Niobium (coming from Columbite) and Tantalum (which is from Tantalite). Coltan is highly appreciated for its properties. It is used to produce the high-performance, small dimension, capacitors for smart phones and mobile ICT devices. While there are severe environmental implications related to the mineral extraction, there are also aspects of its mining that have a clear reference to fair ICT. The mining firms extracting these materials, in the Democratic Republic of Congo in Africa, are suspected of using children as miners and, at the same time, of financing war (OECD 2004; Vazquez-Figueroa 2010).

On only very few occasions have the conditions of workers employed in the various phases of the ICT life cycle been under scrutiny. One of these was at a May 01999 conference in Soesterberg, the Netherlands, where the so-called Soesterberg principles on electronic sustainability commitment were defined: 'Each new generation of technical improvements in electronic products should include parallel and proportional improvements in environmental, health and safety, as well as social justice attributes' (Soesterberg 1999).

7.5.4   Fair ICT and the Health of Workers

Fair ICT should go further than employment rights. It needs to involve health and safety of workers and safe manufacturing systems.

More than 30 years ago, on 17 January 01985, in the United States of America, the *San Jose Mercury News*, a well-known Silicon Valley newspaper, published a front-page headline: 'Birth Defects Study Shows Link to Area of Toxic Spill'. This news was completely out of the ordinary and was considered to be shocking. The high-tech industry was accused of generating toxic hazards throughout its life cycle (from design and production, to consumption and disposal). Vast energy and raw material resources—including highly toxic materials—were acknowledged as being required to produce ICT. Semiconductor workers reported higher illness rates, and women's miscarriage rates were significantly higher than in other industries. These phenomena were strongly related to both the ground and water pollution in the area due to leakage from some chip manufacturing sites (SJMN 1985).

7.5.5   Fair ICT and Its Open Spanning Layers

On another, more positive, front, fair ICT can imply ICT that contributes to the social and economic aspects of society through its innovation potential.

Innovation in ICT is at least in part based on the definition of openly defined layers through which other people can use or reuse systems and technologies. If a specific product is completely closed, so that no one else can develop it or program it, then its innovation potential is restricted. Only the corporate owner of the product (or in many situations the patent's owner) can take advantage of this situation.

Here it is worthwhile introducing the notion of the 'spanning layer' that separates the processes and applications from the underlying infrastructure (Clark 1988, 1994, 1995). Examples can be cited from a range of different industries: for example, transport, software, and hardware. One noteworthy example of a spanning layer is the open standard intermodal freight container used for transportation. Designed in 01956, freight containers have enabled advances in many forms of transport: the same container can be moved around whether on a railway track, a train, or a ship. Its creation and development triggered rapid expansion in the development of the transport industry (Levinson 2008). In ICT, the most well-known spanning layer is the basic protocol of the Internet: TCP/IP (Cerf and Kahn 1974). The open standard of the packets of bits enables people to define new applications on top of the TCP/IP and to develop new channels for transporting bits under it.

After its initial definition in 01974, the Internet and the Web are still flourishing more than 40 years later. This historic open definition of standards is the main trigger for innovation and for social and economic benefits of the Internet.

Since its initial launch some 30 years ago, more examples are increasingly available of the possibilities for fairness offered by open and free software. Open hardware offers similar possibilities. Examples from both these two fields follow.

Free and open software was defined by the Free Software Foundation in 01985 as being related to users' freedom to run, copy, distribute, study, change, and improve software (Stallman 1985). In 02012, it was estimated that the contribution of open source software to the economy of the European Union was around €456 billion a year. This calculation included both the direct cost savings and indirect cost savings. The direct cost savings involved €114 billion in licenses. The indirect cost savings were €342 billion in reduced project failures and increases in productivity and efficiency due to open software's better quality, its lower costs for code maintenance, and its ease of improvement (Hillenius 2012; Daffara 2012).

These forms of software are of particular interest to the public sector, due to the considerable pressures it is under to reduce its costs. The community value of free software is becoming increasingly important for many regional governments that are trying to stimulate a growth in local high-tech companies. These high-tech companies are able to support their clients' organisations by developing new business models. They no longer sell 'proprietary bits', but they provide live consultancy and personalise, customise, and maintain software applications. In Italy, an important example is the local government of Trentino (Trentino 2012). In Germany, another example is the LiMux initiative in the city of Munich. Munich migrated its software systems, including more than 14,000 public employees' personal computers and laptops, to free and open source software (Munich 2012); the first report on the project's economic results estimated a savings of more than €10 million (H-online 2012).

Open hardware as a notion has developed only since the beginning of the twenty-first century. This relatively recent phenomenon (Lahart 2009) is based on hardware design that is open to all, including the schematics, bills of materials, printed and integrated circuit layouts, and the software needed to run the hardware. One of the most famous examples of open hardware is the Arduino platform. Arduino is not just an electronic circuit board, it is also a collection of sensors and add-ons, a repository of free software and configurations for thousands of projects that are ready to use. More importantly, it is a community where several thousands of people exchange knowledge freely (www.arduino.cc).

In Slow Tech, as in Slow Food, one of the important characteristics is of local communities. In Slow Food, this importance is implicit in the origins of crops and animals, the use of recipes that are traditional to a particular region, and the recognition of the micro-ecosystem of which food is a part. The analogies in ICT, and Slow Tech, include stakeholder involvement, participatory design, prototyping, and the decentralisation of production, for example, through three-dimensional printers.

Recent research confirms that open hardware and software are contributing to the launch and growth of smaller local ICT companies (W3Cook 2016).

7.5.6   Fair ICT and War

When investigating the ethics of fair ICT applications, one of the issues that crops up immediately is the deployment of ICT in a war scenario. One of the most profound questions raised by such a reflection is: can computer professionals put their knowledge and intelligence at the service of war? Is it fair to use ICT in a war scenario, to automate killing? These preoccupations are commonly explored by those who work in the computing field. Two examples follow.

An advertisement appeared in the West Coast edition of *The New York Times* on Tuesday, 18 June 01991, sponsored by a major international association of computer professionals dealing with the social and ethical impacts of ICT—Computer Professionals for Social Responsibility (CPSR). It appealed for new priorities and, rather than the smart bombs first used in the armed conflict against Iraq, saw a call for human intelligence (CPSR 1991).

In the 01980s, these kinds of principles and this form of reasoning also underpinned the position of David Parnas, Canadian University Professor, taken against the US government's initiative officially called the Strategic Defense Initiative and more commonly known as Star Wars. His seminal article of 01985, published in the *Communications of the ACM*, begins by identifying the several ways in which the software required at that time by the Strategic Defense Initiative would not be trustworthy (Parnas 1985). Later, in the same article, Parnas described his decision to resign from his university post as being based on human judgement. Hence, in focusing on the limits of machines' reliability—he did not expect research in artificial intelligence to help in building reliable military software—he reflected many of the same concerns as Weizenbaum (1976).

In contemporary terms, in the middle of the second decade of the twenty-first century, there is a simple consequence to applying Parnas' arguments. The outcome should be to request a moratorium on the development of robot warriors, autonomous weapons, or machines with sensors designed for military purposes which are able to take decisions and are designed to kill human beings (Anthony 2015). The position taken by Parnas shows that it is indeed feasible for scientists and designers to take steps away from the work they are asked to undertake, think carefully about the implications of their actions, and refuse to be involved in certain research assignments. Herein lie concerns for responsible, sustainable, and ethical uses of ICT.

REFERENCES

ACM. (2015). *Association for Computing Machinery—Code of Ethics.* https://www.acm.org/about-acm/acm-code-of-ethics-and-professional-conduct. Accessed [9 October 2017].

Anthony, S. (2015, July 27). Musk, Hawking, Wozniak Call for Ban on Autonomous Weapons and Military AI. *Ars Technica UK.*

Barboza, D., & Bradsher, K. (2012, September 24). Foxconn Plant Closed After Riot, Company Says. *New York Times.*

Bateson, G. (1979). *Mind and Nature.* New York: Dutton.

Berridge, C. (2014, April 9). Breathing Room in Monitored Space: The Impact of Passive Monitoring Technology on Privacy in Independent Living. *The Gerontologist.* https://doi.org/10.1093/geront/gnv034.

Bjerknes, G., & Bratteteig, T. (1995). User Participation and Democracy. A Discussion of Scandinavian Research on System Development. *Scandinavian Journal of Information Systems, 7*(1), 73–98.

Bødker, S. (1996). Creating Conditions for Participation: Conflicts and Resources in Systems Design. *Human Computer Interaction, 11*(3), 215–236.

Bradley, G. (2017). *The Good ICT Society. From Theory to Actions.* Abingdon: Routledge Focus.

Bradshaw, K., & Duhigg, C. (2012, December 28). Quietly, Better Work Conditions Take Hold at Chinese Factories. *International Herald Tribune.*

Briggs Myers, I. (1987). *Introduction to Type: A Description of the Theory and Applications of the Myers-Briggs Type Indicator.* Palo Alto: Consulting Psychologists Press.

Burton, G. (2013). Analysis: Does PC Performance Affect Productivity? *Computing.* http://www.computing.co.uk/ctg/analysis/2257838/analysis-does-pc-performance-affect-productivity. Accessed [9 October 2017].

Cabinet Office. (2009). *A Guide to Social Return on Investment.* London: Cabinet Office. https://www.bond.org.uk/data/files/Cabinet_office_A_guide_to_Social_Return_on_Investment.pdf. Accessed [9 October 2017].

CEPIS. (2015). *Council of European Professional Informatics Societies—Ethics.* http://www.cepis.org/index.jsp?p=940&n=2849. Accessed [9 October 2017].

Cerf, V. G., & Kahn, R. E. (1974). A Protocol for Packet Network Intercommunication. *IEEE Transactions on Communications, 22*(5), 637–648.

Chesbrough, H. W. (2015). *New Frontiers in Open Innovation.* Oxford: Oxford University Press.

Cicero. (1998). *De Amicitia (44 BC)* (On Friendship). Bristol: Bristol Classical Press.

Clark, D. (1988). The Design Philosophy of the DARPA Internet Protocols. *Computer Communication Review, 4*, 106–114.

Clark, D. (1994, 1995). *Interoperation, Open Interfaces and Protocol Architecture.* MIT Lab. of Comp. Science (LCS).

Covey, R. S. (1989). *The Seven Habits of Highly Effective People.* New York: Free Press.

Covey, R. S. (1996). *First Things First.* New York: Free Press.

CPSR. (1991). *The CPSR Newsletter,* Spring 1991, Computer Professionals for Social Responsibility. http://cpsr.org/prevsite/publications/newsletters/old/1990s/Winter_Spring1991.txt. Accessed [9 October 2017].

CPSR. (2008). *Computer Professionals for Social Responsibility, Participatory Design.* http://www.cpsr.org/issues/pd. Accessed [9 October 2017].

Crawford, D. L. (2004). *The Role of Aging in Adult Learning: Implications for Instructors in Higher Education.* John Hopkins School of Education. http://www.education.jhu.edu/PD/newhorizons/lifelonglearning/higher-education/implications/. Accessed [9 October 2017].

Creative Commons. (2015). http://creativecommons.org/. Accessed [9 October 2017].

Daffara, C. (2012, September 24). Estimating the Economic Contribution of Open Source Software to the European Economy. In *Open Forum Academy Conference Proceedings,* Brussels.

Davenport, T. H., & Beck, J. C. (2001). *The Attention Economy: Understanding the New Currency of Business.* Boston: Harvard Business School Press.

De Carli, L. (1997). *Internet. Memoria e oblio.* Torino: Bollati Boringhieri.

Duhigg, C., & Barboza, D. (2012, January 25). In China, Human Costs Are Built into an iPad. *New York Times.*

Duhigg, C., & Bradsher, K. (2012, January 23). How the U.S. Lost Out on iPhone Work. *New York Times.*

EGAIS. (2012). *Ethical Governance of Emerging Technologies.* SIS7-CT-2009-230291, 2009–2012. www.egais-project.eu. http://cordis.europa.eu/project/rcn/91156_en.html. Accessed [9 October 2017].

ETICA. (2011). *Ethical Issues of Emerging ICT Applications.* FP7-GA 230318, 2009–2011. www.etica-project.eu. Accessed [9 October 2017].

European Commission. (2002a). Directive 2002/96/EC of the European Parliament and of the Council of 27 January 2003 on *Waste Electrical and Electronic Equipment (WEEE)*—Joint declaration of the European Parliament, the Council and the Commission relating to Article 9, eur-lex.europa.eu. Accessed [9 October 2017].

European Commission. (2002b). Directive 2002/95/EC of the European Parliament and of the Council of 27 January 2003 on the *Restriction of the Use of Certain Hazardous Substances in Electrical and Electronic Equipment (RoHS),* eur-lex.europa.eu. Accessed [9 October 2017].

European Commission (2012a), Responsible Research and Innovation, Research and Innovation, Science in Society, 2012. http://ec.europa.eu/research/science-society/. Accessed [9 October 2017].

European Commission. (2012b). *Digital Agenda for Europe, Futurium.* http://ec.europa.eu/digital-agenda/futurium/. Accessed [9 October 2017].

European Commission. (2012c). FP7-FET Proactive Initiative: *Towards Zero-Power ICT* (2zeroP). http://cordis.europa.eu/fp7/ict/fet-proactive/2zerop_en.html. Accessed [9 October 2017].

European Commission. (2012d). *Ethics Review*, Research and Innovation, Science in Society, 2012. http://ec.europa.eu/research/science-society/. Accessed [9 October 2017].

Fettweis, G., & Zimmermann, E. (2008, September 8–11). ICT Energy Consumption—Trends and Challenges. In *The 11th International Symposium on Wireless Personal Multimedia Communications (WPMC 2008)*, Lapland.

Filler, A. (2009). The History, Development and Impact of Computed Imaging in Neurological Diagnosis and Neurosurgery: CT, MRI, and DTI. *Nature Precedings, 7*, 1–69.

Floridi, L. (2012). Big Data and Their Epistemological Challenge. *Philosophy & Technology, 25*(4), 435–437.

FVG. (2016). *Il Friuli Venezia Giulia partecipa alla conferenza internazionale 'Information and Communication Technologies for Ageing Well and e-health' (ICT4AWE 2016).* http://www.regione.fvg.it. Accessed [9 October 2017].

GeSI. (2015). *#Smarter2030, ICT Solutions for 21st Century Challenges, Global e-Sustainability Initiative*, Brussels. Available at: http://gesi.org/portfolio/project/82. Accessed [9 October 2017].

Goldhaber, M. H. (1997). The Attention Economy on the Net. *First Monday, 2*(4), 3.

Gotterbarn, D. (1992). Software Engineering Ethics. In J. J. Marciniak (Ed.), *Encyclopedia of Software Engineering.* New York: John Wiley & Sons, Inc.

Gotterbarn, D. (2015). The Creation of Facts in the Cloud—A Fiction in the Making. *Computers & Society.* The Newsletter of the ACM Special Interest Group on Computers and Society Special Issue on 20 Years of ETHICOMP, Special Editor(s): M. Coeckelbergh, B. Stahl & C. Flick, (Eds.): Vaibhav Garg and Dee Weikle. http://www.dmu.ac.uk/documents/research-documents/technology/ccsr/20-years-of-ethicomp-si.pdf. Accessed [9 October 2017].

Greenpeace. (2012). *Guide to Greener Electronics.* November 2012 Release. www.greenpeace.org. Accessed [9 October 2017].

H-online. (2012). *Linux Brings over €10 Million Savings for Munich.* http://www.h-online.com/open/news/item/Linux-brings-over-EUR10-million-savings-for-Munich-1755802.html. Last modified 23 Nov 2012. Accessed [9 October 2017].

Habermas, J. (1996). Kampf um Anerkennung im demokratischen Rechtsstaat. In *Die Einbeziehung des Anderen. Studien zur politischen Theorie* (pp. 237–276). Frankfurt am Main: Suhrkamp.

Hillenius, G. (2012). *Contribution of Open Source to Europe's Economy: 450 Billion Per Year.* https://joinup.ec.europa.eu/news/contribution-open-source-europeseconomy-450-billion-year. Last modified 11 Oct 2012. Accessed [9 October 2017].

ISF. (2015). Informatici Senza Frontiere, *Progetto ABC Computer*, Trento. www. informaticisenzafrontiere.org. Accessed [9 October 2017].

Johnson, D. G. (2009) *Computer Ethics* (4th ed.). Saddle River: Pearson International Edition, Prentice Hall.

Jonas, H. (1985). *The Imperative of Responsibility: In Search of an Ethics for the Technological Age* (H. Jonas & D. Herr, Trans.). Chicago: University of Chicago Press. Originally published as *Das Prinzip Verantwortung: Versuch einer Ethik fur die technologische Zivilisation* [Frankfurt am Main: Insel Verlag, 1979].

Kanamaru, N., Tezuka, H., Fukayama, A., Nakamura, Y., Yamaguchi, H., & Motegi, M. (2015, July). Creating *Omotenashi* Services for Visitors and Spectators in 2020. *NTT Technical Review, 13*(7), 1–5. https://www.ntt-review.jp/archive/ntttechnical.php?contents=ntr201507fa6.html. Accessed [9 October 2017].

Kavathatzopoulos, I. (2011). The Relation Between ICT, Sustainability and Ethics. In D. Whitehouse, L. Hilty, N. Patrignani & M. Van Lieshout (Eds.), *Social Accountability and Sustainability in the Information Society: Perspectives on Long-Term Responsibility,* special issue of *Notizie di Politeia, 27*(104), 79–90.

Lahart, J. (2009, November 27). Taking an Open-Source Approach to Hardware. *The Wall Street Journal.*

Lennefors, T. T. (2013, June 12–14). Marketing the Information Society: Sustainability, Speed and Technomass. In T. W. Bynum, W. Fleishman, A. Gerdes, G. Moldrup Nielsen & S. Rogerson (Eds.), *The Possibilities of Ethical ICT,* ETHICOMP 2013 Conference Proceedings (pp. 310–315). Available at https://www.researchgate.net/publication/260298598_ ETHICOMP_2013_Conference_Proceedings_The_possibilities_of_ethical_ ICT. Accessed [9 October 2017].

Levinson, M. (2008). *The Box: How the Shipping Container Made the World Smaller and the World Economy Bigger.* Princeton: Princeton University Press.

Margolis, J. (2001). *A Brief History of Tomorrow: The Future Past and Present.* London: Bloomsbury Publishing.

McGinn, D. (2011, May). Being More Productive. *Harvard Business Review.* Available at https://hbr.org/2011/05/being-more-productive. Accessed [9 October 2017].

MIT. (2012). Sustainability Nears a Tipping Point. *MIT Sloan Management Review,* Winter 2012 Report.

Moore, G. (1965). Cramming More Components onto Integrated Circuits. *Electronics Magazine, 38*(8), 114–117.

Munich. (2012). *Das Projekt LiMux.* www.muenchen.de. Accessed [9 October 2017].

OECD. (2004). *Illegal Exploitation of Natural Resources in the Democratic Republic of Congo: Public Statement by CIME*. www.oecd.org.

OECD. (2015). *Students, Computers and Learning: Making the Connection*. PISA, OECD Publishing. http://www.oecd-ilibrary.org/education/students-computers-and-learning_9789264239555-en. Accessed [9 October 2017].

Palfrey, J., & Gasser, U. (2008). *Born Digital, Understanding the First Generation of Digital Natives*. New York: Basic Books.

Parnas, D. L. (1985). Software Aspects of Strategic Defense Systems. *Communications of the ACM, 28*(12).

Patrignani, N. (2009a, June). Un'etica per i robot. *Wired* (Italian edition), p. 24.

Patrignani, N. (2009b, March 17–18). *The Map Is Not the Territory. Applying Wiener's (and Bateson's) Ethical Lesson to Future Computing and Brain Imaging Applications*. Brains in Dialogue on Brain Imaging Conference, Clare College, Cambridge.

Patrignani, N., & Whitehouse, D. (2015). The Clean Side of Slow Tech: An Overview. *Journal of Information, Communication and Ethics in Society, 13*(1), 3–12.

Patrignani, N., Laaksoharju, M., & Kavathatzopoulos, I. (2011). Challenging the Pursuit of Moore's Law: ICT Sustainability in the Cloud Computing Era. In D. Whitehouse, L. Hilty, N. Patrignani & M. Van Lieshout (Eds.), *Social Accountability and Sustainability in the Information Society: Perspectives on Long-Term Responsibility*, special issue of *Notizie di Politeia, 27*(104).

Petrini, C. (2011). *Buono, Pulito e Giusto*. Torino: Einaudi.

Peugeot. (2015). *Peugeot Concours Design*. https://coolcarsdesign.wordpress.com/category/peugeot-design-contest/. Accessed [9 October 2017].

Prystay, C. (2004, September 23). Recycling E-waste. *Wall Street Journal*.

Ressler, C., & Thompson, J. (2008). *Why Work Sucks and How to Fix It*, Amazon Media.

Rodotà, S. (2005). *Intervista su privacy e libertà*. Bari: Laterza Editore.

Rogerson, S., & Gotterbarn, D. (1998). The Ethics of Software Project Management. In G. Colleste (Ed.), *Ethics and Information Technology*. Delhi: New Academic Publisher.

SHRM. (2012, July 12). Survey Findings: Work/Life Balance Policies. *Society for Human Resource Management*. http://www.shrm.org

Sigismund Huff, A., Möslein, K. M., & Reichwald, R. (2015). *Leading Open Innovation*. Cambridge, MA: MIT Press.

SJMN. (1985, January 17). High Birth Defects Rate in Spill Area. *San Jose Mercury News*.

Soesterberg. (1999, May 16). Declaration Adopted by *Trans-Atlantic Network for Clean Production Meeting*, Soesterberg.

Spector, J. M., Ifenthaler, D., & Kinshuk, P. (2014). *Learning and Instruction in the Digital Age*. New York: Springer.

Stallman, R. (1985). The GNU Manifesto. *Dr. Dobb's Journal of Software Tools, 10*(3), 30.

Trentino. (2012, July 17). *Software libero e open source sono nell'ordinamento provinciale.* Consiglio della Provincia Autonoma di Trento. www.consiglio.provincia.tn.it. Accessed [9 October 2017].

Van Abel, B., Evers, L., Klaassen, R., & Troxler, P. (2011). *Open Design Now: Why Design Cannot Remain Exclusive.* Amsterdam: BIS Publishers.

Vazquez-Figueroa, A. (2010). *Coltan.* Barcelona: Ediciones B.

Von Hippel, E. (2006). *Democratizing Innovation.* Cambridge, MA/London: MIT Press.

W3Cook. (2016). OS Market Share and Usage Trends. *Recipes of the WWW.* http://www.w3cook.com/os/summary. Accessed [9 October 2017].

Weizenbaum, J. (1976). *Computer Power and Human Reason: From Judgment to Calculation.* New York/San Francisco: W.H. Freeman.

Whitehouse, D., Burmeister, O., Duquenoy, P., Gotterbarn, D., Kimppa, K., Kreps, D., & Patrignani, N. (2015, September). Twenty-Five Years of ICT and Society: Codes of Ethics and Cloud Computing. In M. Coeckelbergh, B. Stahl & C. Flick (Eds.), *20 Years of ETHICOMP*, a special issue of V. Garg & D. Weikle (Eds.), *Computers & Society, The Newsletter of the ACM Special Interest Group on Computers and Society, 45*(3), 18–24.

CHAPTER 8

Applying Slow Tech in Real Life

Abstract This chapter describes Slow Tech ideas that are directed mainly towards young engineers: it investigates what are some of the main influences on the professionals who are developing information technology. It is a chapter that can be of special support to industrialists and people working in the commercial sector.

Today's greater awareness of complexity is contrasted with the engineering focus on reductionism. In complete contrast to the reductionist approach, several new initiatives may instead be helpful to young people who are launching their careers at the start of the twenty-first century—particularly, the approach of responsible research and innovation and computer ethics. Currently, a Slow Tech approach to information technology is needed, an approach that recognises the complexity of multiple, interconnected systems. In this chapter examples of applications and real case studies that illustrate the Slow Tech approach are provided. The first of these is the new technology of cloud computing. The cases that follow illustrate the progress of three different companies that have each, in its own way, applied Slow Tech ideas, over several decades since the 01960s but more prominently during the last decade: they are Olivetti and Loccioni from Italy and Fairphone from the Netherlands. The fact that these case studies come from real companies implies the start of a discussion with several groups of stakeholders.

N. Patrignani, D. Whitehouse, *Slow Tech and ICT*,
https://doi.org/10.1007/978-3-319-68944-9_8

Keywords Cloud computing • Companies • Complexity • Computer professionals • Dialogue • Engineers • Fairphone • Loccioni • Methods • Olivetti • Reductionism • Stakeholders • Users • Young generations

In this chapter, we examine the several stakeholders involved with the introduction of Slow Tech as a compass for finding new routes in ICT. We examine how one might use Slow Tech to examine a current or a future computing challenge—in this case, cloud computing. Then we provide some case studies of real companies that can be described as Slow Tech in their orientation.

Specific challenges face each of these groups of stakeholders; in particular, one is for computer professionals: complexity.

8.1   Moving Towards an Understanding of Complex Human Beings

There are a lot of changes that are now taking place in all the sciences. The traditional reductionist approach is evolving towards a range of new approaches that take complexity into account: this is the complex systems approach (Science 2009). Going beyond the reductionism so prevalent during the last centuries means recognising that over-simplification does not function well any more.

Seeing ICT systems as separate from humanity—without taking into account either the context in which they will be used or, indeed, real users—results from applying a reductionist approach to ICT.

Engineers and designers should not concentrate on ICT without examining the background, context, and purposes to that technology. They could instead start asking actual end users about what it is that they need from ICT. In digital health applications, for example, it is fundamentally important to see a more complete picture of the whole complexity of nations' health, well-being, and welfare systems.

The current level of complexity in society requires a new paradigm for ICT: a paradigm that includes human beings in the design of ICT systems.

Of course, human beings introduce greater complexity into any system. They share certain characteristics in common: purely as examples, they forget, they become distracted, and they grow old. If human beings are included in the design process, the main properties of a system will emerge even while the actual system is being designed. If the people who are going to use the ICT system are involved directly in its design, it is likely that all concerned will start seeing beyond the technology. Maybe, as a result, many properties of the system will be perceived before deployment or in advance of the system going live.

For computer professionals, this shift away from reductionism to complexity accompanied by an embedding of values into design opens up two opportunities. First, engineers can become socio-technical analysts who will be more effective personnel able to explore and enjoy the richness of ICT. Second, as the future designers of ICT systems, they can learn to design the world's artefacts by taking into account human beings and their limits as well as the limits of the planet.

As soon as human beings are included in the design picture, certain new competences will become important requirements for engineers. Examples include an awareness and application of Value Sensitive Design (Friedman 1996; Nissenbaum 1998) and the principles of Human Computer Interaction (Benyon 2010).

What therefore are the values that will drive ICT design? On what values can people agree? What skills, competences, and processes will be needed by these future designers?

8.2   STARTING A DIALOGUE WITH STAKEHOLDERS

It is crucial to begin a dialogue about Slow Tech with several audiences and stakeholders. These stakeholders include—but ultimately are not limited to—the following four categories of people and organisations: the younger generations, end users or consumers, ICT companies, and computer professionals. The complexity and challenges facing computer professionals are explored in depth.

A description is offered for each of the types of dialogues that could be held with these four types of stakeholders: these discussions are outlined in different depths and intensities. Particularly the questions that could be raised in and with companies are those which are explored here most intensively.

Of course, similar proposals could be made to many other kinds of stakeholders involved in ICT design, development, and use; however, they are not covered here.

8.2.1   *Younger Generations*

It is important to focus on young people—the millennials—who have grown up with technology and may be more open to changes in technology and society (Rogerson 2015), especially those who are developing solid, critical thinking about their future lifestyles and jobs. As astute citizens and consumers, today's young people are starting to develop an ethics for the future, they are selecting carefully the food they eat and where and how they travel. Now some of them are beginning to ask questions of the individuals and groups involved in the domain of the social and ethical issues of computing: What are the correct choices to make about ICT? What are the decision criteria to apply when someone has to develop, buy, use, and dispose of ICT systems? What should people do if they want to be fair, democratic, and egalitarian when they buy and use a computer or a smartphone? What about taking into account the relevant social and environmental issues? If it is so important to choose food carefully, and focus on vegetables that are grown locally (e.g. that have travelled zero kilometres or close to that distance and are grown by a local farmer or producer or even in one's own garden, allotment, or plot), why should one not also look for a zero-kilometre computer, by starting to reuse and recycle computers?

8.2.2   *Users*

Users and consumers are increasingly becoming critical consumers. They want to know the implications of their choices. There are at least three ways in which they are examining their choices and acquisitions. One is that they are starting to question where technology comes from (i.e. they want to know about the living conditions of workers who produce ICT) and are insisting on the sustainability of ICT in the long term (Barboza and Bradsher 2012). This kind of critical thinking on the part of contemporary consumers means that people are beginning to look carefully at the entire life cycle of the products that they buy and use. This is a form of fair trade in which the fairness of a commercial transaction can be applied not just to bananas or coffee but to computers, smartphones, cloud services. Two, they now want products that are repairable or recyclable or that have only a very light impact on the environment and are not harmful to the planet. Three, users sometimes feel that they are slaves of their ICT: their use of it is out of kilter or imbalanced. Instead, people are beginning to feel that they need forms of ICT that can really help them (e.g. digital health), that bring them a new feeling of balance in their lives or offer them a rebalancing between business time and free time. They are seeking to revert to slower and more traditional activities: cooking, drawing, gardening, knitting or sewing, and putting down or away their technical devices or gadgets.

8.2.3   Computer Professionals

It is important to start a dialogue with computer professionals, the people who are responsible for the design and development of ICT. When engineers simply look at the technology alone, their narrow search will offer them very little insight into the system's important properties.

This dialogue and interaction can be looked at from several perspectives: in the early stages of career development when high school students and university students first move into computing; throughout the whole employment process; or from a more generalised, overall view of science and engineering. From an educational point of view, one should focus on the way in which students approach technologies and start their engineering or computing studies.

The typical engineering approach is a problem-solving one in which young people concentrate just on the technological challenges. What are young people's motivations for choosing to go to an engineering school? In many educational institutions and universities, young engineers make this choice because they want to escape from the complexity of having to deal with other human beings; they do not like the intricacies of the humanities; instead, they want to prioritise a reductionist approach to the world and simplify reality through the development of simple models. They then start working on these overly simple models. This rather simplistic approach often drives a technological push, as a result of which technology is seen as looking for problems to solve and providing their solutions.

It is scarcely surprising that computer engineers are often positioned at an extreme pole in terms of the reductionist approach, due either to personal background or history (Weinberg 1971). In general, computer scientists do not want to be involved in or with complexity. They tend to want to live in a virtual space that is limited to their keyboard and their screen. They risk escaping from reality and entering their own virtual reality or artificial world (Silberman 2001). As a result, a typical engineering response is to consider using technology to improve human performance rather than to respect or rehabilitate human behaviour, perhaps with a focus on transhumanism (details of the transhumanist approach are located in De Grey and Rae 2007; Garreau 2005; Hughes 2004; Kurzweil 2006). (The transhumanist approach minimises the human side of the convergence between technology and people: it switches off the human aspects and turns on the machine aspects.)

In the field of engineering, engineers are accustomed to using the language of science and reasoning in terms of conditions that are true and false. However, instead, when dealing with human values, it is appropriate to reason in terms of right or wrong: the language of ethics. Such a discussion is much more complicated than an engineering-related conversation because it implies being aware of and open to other people's values and desires. As a designer, it is important to be able to interact with the very people who will use the computer systems that one is helping to design.

Questions of right and wrong about computers, and the search for a reasoned approach to these questions, are implicit in the field of applied ethics that is known as computer ethics (Johnson 2009). (See Chap. 6 for more details.) For a period of at least 25 years, computer ethics has been a discipline in many universities in the United States of America and around the world (ACM 1991) but it is still an unknown discipline in computer engineering schools in many countries.

8.2.4   ICT Companies

It is vital to start a dialogue with real ICT companies and organisations. What should corporations do when they too have to make choices about ICT? What would be a wise use of technology? How is it possible to introduce ICT into companies so that it really makes a difference to them? There is a whole series of dilemmas that need to be explored and answered (Dilemmas 2015). The dilemmas include decisions to be made at the level of the employee, employee procedures and guidelines, technology design, the ICT value chain, the availability of international standards, and the definition of appropriate corporate social responsibility procedures. For example, for both employers and employees, when dealing with electronic communications, what is a wise way to use email? Does a person really need to be 'always-on'? Would it be appropriate to encourage people not to check their emails constantly? Or for people to disconnect themselves from the net or Web either for at least four hours a day or 50 per cent of their working time? Since many companies now employ knowledge workers—people whose work involves handling information—they should consider the quality of life of these employees and pose themselves questions about how they can attract and retain these workers. Companies should also be aware that ICT, by itself, does not always improve the quality of knowledge or the productivity of knowledge workers. Corporations and enterprises have to prepare a good environment for their knowledge workers, by providing a workplace that is hospitable.

In either their design or their use of technologies, many companies could make a Slow Tech difference by introducing recyclable-by-design concepts (McDonough and Braungart 2002). By using standard interfaces and interchangeable modules in their equipment, companies could contribute to addressing the challenge of reducing e-waste (Carrol 2008). Minimising the use of rare materials would be another positive initiative to be made in the direction of creating sustainable ICT. In terms of the ICT value chain, many ICT companies now rely on external business partners for the production of ICT devices. Such ICT companies do not control the entire ICT value chain. They are, nevertheless, seen increasingly as also having a responsibility for the working conditions of employees based on the premises of their technology providers—even when these are located on other continents (Duhigg and Bradsher 2012; Pouille 2012). Companies can therefore start demanding, of themselves and consequently of their providers, that they respect appropriate international standards. Examples of suitable standards include the *United Nations Global Compact* (UNGC 2015), the *Global Reporting Initiative* (GRI 2015), and the guidance provided by the *ISO 26000 on Social Responsibility* (ISO 2015).

It is now high time for the ICT industry to start seriously defining a strategy for corporate social responsibility (Patrignani and Whitehouse 2014, 2015).

8.3   Applying the Slow Tech Method

The Slow Tech approach offers a simple tool for analysing complex scenarios. It can be difficult to find simple yes and no answers to Slow Tech questions when they are applied to a specific case. Nevertheless, Slow Tech provides some useful questions that can stimulate reflections about the co-shaping of the future that may take place between ICT and human beings. A good example to examine in this case is cloud computing.

For a complete application of the Slow Tech approach to this real scenario, the very same questions should be posed recursively from the point of view of the cloud providers to their ICT vendors suppliers and so on.

Of course if all stakeholders apply this Slow Tech approach, then we can hope that the entire stakeholders network will show an emergent property: it can progressively become more and more good, clean, and fair.

The cloud provides a global infrastructure that has the following characteristics: it is a network based on broadband, with computing servers that act as shared platforms. The cloud is typified by resource pooling and multi-tenancy, rapid scalability and elasticity, and measured/metered services for billing purposes. It is available on demand, that is, it is self-service.

The cloud looks different, and has quite different advantages and disadvantages, depending on what stakeholder perspective it is viewed from.

From the perspective of a small- or medium-sized enterprise (i.e. in the view of many small firms), this option can offer a number of benefits. The firm can begin to access different kinds of software application from a remote location (known as Software as a Service or SaaS), any kind of development environment (Platform as a Service, PaaS), and any kind of ICT resource online (Infrastructure as a Service, IaaS). The organisation does not need to have dedicated computer facilities, it just requires a network and devices with browsers.

From another perspective, end users risk losing control of their computational and storage capabilities completely. Users begin to be attracted towards a superficial kind of computing where the only physical equipment needed is a touchscreen that navigates the surfaces of clouds.

What steps are needed to assess the Slow Tech aspects of cloud computing? The first step is to build the stakeholders' network (see Fig. 8.1). In this example, based on a simplified network, one can concentrate on the main actors only: cloud providers and cloud users. It can be assumed that the users simply subscribe to a form of Software as a Service.

Applying a Slow Tech checklist of questions, from the point of view of cloud users, would mean that users start to ask the cloud provider a series of important questions related to the service subscribed.

As a bare minimum, this simply means asking whether the cloud computing concerned is good, clean, and fair. Under these three categories, other sub-questions can be posed, such as, asking the cloud provider whether it develops its software applications by using a user-centric approach. More details follow.

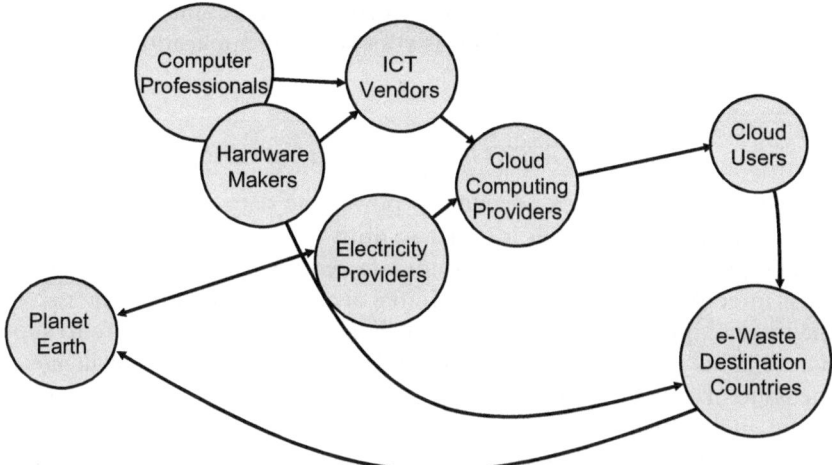

Fig. 8.1   Cloud computing stakeholders' network

8.3.1   Is This Cloud Computing Good?

It is only rarely that human beings are included in the design of ICT environments. Cloud computing is very much technology push. Since its dawn, the cloud has been based on the availability of the immense storage and processing capabilities held in gigantic data centres around the globe. Since the titans of the Web initially used only around 20 per cent of their Web-based capabilities, they determined that their unused ICT resources could be sold online, with the intention of optimising their investments.

There are many broadly ethical issues that need to be addressed when exploring cloud computing; among them are the borders of control and responsibility, the risk of monopoly and lock in (Sterling 2012), the question of the ownership of the data, the problem of many hands touching the data (who is doing what, and when, on the machines in the cloud?), and function creep (improper use of users' data) (ETICA 2011).

It is, however, worthwhile saying that, for many users, cloud computing is nevertheless a desirable ICT: they find the simple interface of just a browser and a touchscreen very attractive.

To determine whether a particular form of cloud computing is good, one could ask the cloud provider: is your software application developed by using a user-centric approach?

8.3.2   *Is This Cloud Computing Clean?*

Where does the electricity needed to power the data centre come from? The sustainability of clouds is under scrutiny.

The power consumption of cloud data centres is currently one of the most interesting areas of research. Initial analyses demonstrate the risk of the increasingly large energy sources needed by the infrastructure to support billions of computing requests (Glanz 2012). Another challenge is the issue of e-waste: where does the hardware used in any data centre go at the end of its life? While it can be hoped that the large-scale, concentrated infrastructures needed by the cloud will reduce the risk of the uncontrolled traffic of e-waste, this cannot yet be guaranteed.

8.3.3   *Is This Cloud Computing Fair?*

Where are these data centres based? What kinds of contracts exist with the computer professionals running the data centres? Will the rights of workers be respected?

The stakeholders' network of the ICT landscape is becoming more complicated with the birth of cloud-brokers, for example, these are organisations that do not own any infrastructure, but just buy and resell ICT resources. Thus, the ICT value chain is becoming longer than ever before and more difficult to make transparent.

There is a high risk that a shift will take place in which ICT infrastructures will be moved to countries where salaries are lower and there are fewer workers' rights.

8.4   LOCATING EXISTING EXAMPLES OF SLOW TECH COMPANIES

It is also useful to illustrate these kinds of dilemmas by choosing some current examples of companies that use a Slow Tech approach.

The use of case studies such as these can be particularly appropriate in terms of helping fellow industrialists to understand what is good practice, to identify particular model behaviour and approaches, and—for learning purposes—to identify ways of handling special challenges.

These cases cover technologies such as computers, robots, and mobile telephony. The first of these case studies is an historical one. At least two of them consider in depth not only their manufacturing methods but also the physical environments in which they are themselves located.

8.4.1   Olivetti

This company is related to Slow Tech for its willingness to place human beings centre stage.

Olivetti was a company founded by one of the most important Italian industrialists and visionaries of the twentieth century: Adriano Olivetti (Ivrea 1901–1960). Olivetti was able to combine advanced technologies, innovation, production, profit, solidarity, social responsibility, and beauty in a joint enterprise (Olivetti 1959).

As a fundamental focus, the company shifted its orientation in the mid-twentieth century from typewriters to computers. Among Olivetti's many achievements were the first transistor-based 'mainframe', the Olivetti Elea 9003 in 01959; the first personal computer, the Olivetti P101 in 01965 (Wall Street Journal 1965); and the Olivetti Electronic Center building, located between Torino and Milano, in Italy, designed by Swiss architect, Le Corbusier.

The innovative climate in the Olivetti company was based on the idea of the design and construction of a community (Olivetti 1959). This vision led to one of the most socially advanced working environments of the 01950s. The Olivetti campus included a library, a nursery and schools, houses built for workers, and a development plan intended for the whole city where it was located.

In the area of ICT, Olivetti was one of the main and advanced ICT providers worldwide. By applying Slow Tech principles, one can examine whether the ICT that Olivetti produced was indeed good, clean, and fair. It can be demonstrated that the ICT was indeed good ICT, designed with a human-centric approach. The clean side of the equation is probably the most difficult issue to resolve as the founder of the firm, Adriano Olivetti, died in 01960, well before many environmental reflections began. However, the ICT was certainly a fair form of ICT since the working conditions of Olivetti's employees were among the best in the world at that era. For example, the company introduced nine months of paid maternity leave for its female employees.

According to the vision of Adriano Olivetti, the beauty of the physical workspace and living spaces of the Olivetti company employees was particularly important: therefore, the company is not only a place of production but is also the main engine of economic and social development. As such, it has a definite responsibility both to the community and the country (in this case, Italy) (Archivio Storico Olivetti 2008: 1).

8.4.2   Loccioni

Olivetti's vision of the workplace and the surrounding community is more contemporarily paralleled by that of Loccioni, a high-tech company that develops solutions and integration projects, in central Italy (Patrignani and Whitehouse 2014, 2015). This company is related to Slow Tech for its strong commitment to community values and the environment.

Loccioni can be called a 'technology tailor'. It provides a living example of how a company and territory can co-shape each other by taking into account sustainability, fairness, and beauty (Varvelli and Varvelli 2014). Two of the many Loccioni projects and solutions that can be mentioned in this regard are the Leaf House and APOTECA.

The Leaf House is a carbon neutral house where energy is produced entirely through renewable sources without CO_2 emissions. It is a technologically innovative house: its characteristics of cheapness, simplicity, efficiency, and silence are combined and integrated to create a house made for the environment. It provides an example of saving and respect. It is a clean energy laboratory, a place to be studied and visited so as to awaken people's awareness and educate them about the future. It is a house composed of six flats: indeed, it is a real house where actual people live (Leaf Community 2017).

APOTECA is a robotic solution built by the Loccioni Group. APOTECA is the name from the Ancient Greek or Latin for a storehouse. Today, the term is used to represent a pharmacy. In Loccioni's sense, it refers to a robotic application for hospital pharmacies that addresses a sensitive problem that they face of enabling 'a perfect integration between the department and the oncological pharmacy service' (Loccioni-Humancare 2017: 1).

The APOTECA system is used in many hospitals around the world. The APOTECA stakeholders' network itself is quite complex, since it includes patients and their relatives, nurses, doctors, pharmacy personnel, the hospital organisation, pharmaceutical providers, the technology provider, and many more. In applying the Slow Tech approach to

APOTECA, the main relationship to be focused on is that between the hospital and the ICT provider.

In Loccioni's case, the ICT involved in APOTECA is good: it has been designed and tested with the contribution and participation of the main involved stakeholders (i.e. patients and nurses). Its robotic arm is able to prepare the precise pharmaceutical dosages needed for cancer patients with a high level of safety for the people working in the pharmacy. It is also a clean ICT since the design and production executed by the Loccioni Group is that of a well-recognised leader in environmental care and the development of environmentally sustainable solutions. The ICT is fair because the ICT provider has demonstrated for several years already a strong corporate social responsibility strategy with respect to its employees.

8.4.3   *Fairphone*

This company is related to Slow Tech since it is one of the first examples of commitment to social and environmental care when developing a 'smartphone with social values'.

Fairphone is a social enterprise, that was established in 02010 in Amsterdam, the Netherlands. Its goal was to build a movement for fairer electronics. It started from a smart phone called a Fairphone.

The enterprise has taken a strongly transparent approach towards the phone's supply chain in order to explain the connections between people and products. It addresses the entire value chain, from each of the perspectives of mining, design, manufacturing, and life cycle. It has adopted an ethical approach to the ICT industry throughout all the phases of the phone's life cycle:

- mining (using materials that support local economies in, and minerals coming from, conflict-free countries);
- design (emphasising longevity and reparability, so it is easy to open up and repair the most commonly broken parts of the phone; supplying open source repairing guides; and designing elements based on open software principles);
- manufacturing (by ensuring fair working conditions to workers in factories and working closely with manufacturers that want to invest in employee well-being);
- life cycle (by maximising use, reuse, and safe recycling);
- social entrepreneurship (focusing on social values and transparency towards consumers: breakdown of all costs is provided).

The Fairphone company has already sold more than 60,000 Fairphones and is supported by 39 employees, of 20 different nationalities (Fairphone 2015). Even if 100 per cent fairness, in terms of the treatment of its own employees, is to be proved impossible, for Fairphone as a company, the adoption of each of these principles is definitely a step in the appropriate direction. The company therefore provides a good contemporary example of a Slow Tech approach.

REFERENCES

ACM. (1991). Computing Curricula, Social, Ethical and Professional Issues. *Communications of the ACM, 34*(6), 69–84.

Archivio Storico Olivetti. (2008). Olivetti. Storia di un'impresa. Una iniziativa dell'Associazione Archivio Storico Olivetti. http://www.storiaolivetti.it. Accessed [9 October 2017].

Barboza, D., & Bradsher, K. (2012, September 24). Foxconn Plant Closed After Riot, Company Says. *New York Times.*

Benyon, D. (2010). *Designing Interactive Systems. A Comprehensive Guide to Human Computer Interaction and Interaction Design* (2nd ed.). Harlow: Pearson Education Ltd.

Carrol, C. (2008, January). High-Tech Trash. *National Geographic.*

De Grey, A. D. N. J., & Rae, M. (2007). *Ending Aging: The Rejuvenation Breakthroughs that Could Reverse Human Ageing in Our Lifetime.* New York: St. Martin's Press.

Dilemmas. (2015, September 9–11). *Dilemmas 2015, A Markus Wallenberg Symposium.* Linnaeus University. http://lnu.se/about-lnu/conferences/previous-conferences/2015/dilemmas-2015?l=en. Accessed [9 October 2017].

Duhigg, C., & Bradsher, K. (2012, January 23). How the U.S. Lost Out on iPhone Work. *New York Times.*

ETICA. (2011). *Ethical Issues of Emerging ICT Applications,* FP7-GA 230318, 2009–2011. www.etica-project.eu. Accessed [9 October 2017].

Fairphone. (2015). *Fairphone Fact Sheet.* Fairphone, Amsterdam. Available at: http://www.fairphone.com/wp-content/uploads/2014/07/Fairphone-fact-sheet.pdf. Accessed [9 October 2017].

Friedman, B. (1996, November/December). Value Sensitive Design, *Interactions.*

Garreau, J. (2005). *Radical Evolution. The Promise and Peril of Enhancing Our Minds, Our Bodies—And What It Means to Be Human.* New York: Broadway Books.

Glanz, J. (2012, September 22). Power, Pollution and the Internet. *New York Times.*

GRI. (2015). *Global Reporting Initiative.* www.globalreporting.org. Accessed [9 October 2017].

Hughes, J. (2004). *Citizen Cyborg: Why Democratic Societies Must Respond to the Redesigned Human of the Future.* Boulder: Westview Press.

ISO. (2015). www.iso.org/iso/home/standards/iso26000. Accessed [9 October 2017].

Johnson, D. G. (2009). *Computer Ethics* (4th ed., pp. 13–20). Saddle River: Pearson International Edition, Prentice Hall.

Kurzweil, R. (2006). *The Singularity Is Near: When Humans Transcend Biology.* London: Penguin Books.

Leaf Community. (2017). *Leaf House.* Available at http://www.leafcommunity. com/en. Accessed [9 October 2017].

Loccioni-Humancare. (2017). *Apoteca.* Available at: http://humancare.loccioni. com/about-us/projects/apoteca/. Accessed [9 October 2017].

McDonough, W., & Braungart, M. (2002). *Cradle to Cradle: Remaking the Way We Make Things.* New York: North Point Press.

Nissenbaum, H. (1998). Values in the Design of Computer Systems. *Computers in Society, 28,* 38–39.

Olivetti, A. (1959). *Città dell'uomo.* Milano: Edizioni di Comunita'.

Patrignani, N., & Whitehouse, D. (2014, July 30–August 1). Slow Tech: The Bridge Between Computer Ethics and Business Ethics. In K. Kimppa, D. Whitehouse, T. Kuusela & J. Phahlamohlaka (Eds.), *ICT and Society,* 11th IFIP TC 9 International Conference on Human Choice and Computers, HCC11 2014, Turku, Proceedings. Series: IFIP Advances in Information and Communication Technology (Vol. 431), Springer, ISBN 9783-662-44207-4.

Patrignani, N., & Whitehouse, D. (2015). Slow Tech: Bridging Computer Ethics and Business Ethics. *Information Technology & People, 28*(4), 775–789.

Pouille, J. (2012, Juin). En Chine, la vie selon Apple. *Le Monde diplomatique,* pp. 1, 20–21.

Rogerson, S. (2015). Future Vision. *Journal of Information, Communication and Ethics in Society, 13*(3/4), 346–360.

Science. (2009, July 24). *Complex Systems and Networks,* Special issue, *325*(5939), 357–504.

Silberman, S. (2001, December). The Geek Syndrome. *Wired, 9*(12), 174–183.

Sterling, B. (2012, December 27). Why It Stopped Making Sense to Talk About 'The Internet' in 2012. *The Atlantic.*

UNGC. (2015). *United Nations Global Compact.* www.unglobalcompact.org. Accessed [9 October 2017].

Varvelli, M. L., & Varvelli, R. (2014). *2 Km di futuro: l'impresa di seminare bellezza, Il Sole 24 Ore,* Milan.

Wall Street Journal. (1965, October 15). Desk-Top Size Computer Is Being Sold by Olivetti for First Time in US.

Weinberg, G. M. (1971). *The Psychology of Computer Programming.* New York: Dorset House Publishing.

CHAPTER 9

Energy, Time, and Information: A Long-Term View of ICT

Abstract The book comes to a close with a final chapter. This chapter takes a stance on the planetary and human limits to computing. It takes a look back over the historical developments in technology in general as it has helped to support human life, with an emphasis on the ideas of energy researcher and environmentalist, Daniel Spreng. In historical terms, technology helped human beings in terms of its use of energy to manage processes. Later, information technology, that is, computers used information. Now, just at the period when human beings have reached the limits of energy and information, they discover the good news that there is another dimension—there is time. People can slow down and adopt a Slow Tech approach! With such a responsible, sustainable, and ethical approach to ICT, the future focus needs to be on human happiness and people's human interactions with each other in their communities and societies.

Keywords Brakes • Climate change • Ecosystem • Ethical • Environment • Green IT • Happiness • History • Humankind • Humanity • Interaction • Processes • Spreng • Responsible • Revolution • Sustainable • Sustainability • Tomorrow • Triangle

N. Patrignani, D. Whitehouse, *Slow Tech and ICT*, https://doi.org/10.1007/978-3-319-68944-9_9

This book is based on the importance of limits. In terms of future sustainability, it is important to bear in mind these limits, whether in the two fields of environmentalism or ICT. It is proposed to concentrate on environmental limits since it is these which should inform people's actions, including ICT systems design, for the next 30 years.

While the message about the limits to growth launched by Club of Rome in 01972 (Meadows et al. 1972) fell out of favour during the 01990s and 02000s, it now looks as though it is being much more well received by society as a whole. See, for example, the May 02015 papal encyclical on climate change (Franciscus 2015; Vidal 2015)—especially its third chapter on technological paradigms—or the agenda of the United Nations meeting which was held in Paris in December 02015 (UN 2015).

Thanks to the work of the Intergovernmental Panel on Climate Change (IPCC), the problem of climate change is now firmly on the agenda of policy makers (Elliott 2015). Given the global risk that can arise from the development and potential use of atomic weapons, humanity is beginning to learn constructively from the writings of philosophers (Jonas 1985), futurists (Skrimshire 2010), and researchers about the need for a profound rethinking of technological development and its consequences for current and future generations (Di Paola and Pellegrino 2014). These messages are also taken up in the popular media, with the launch of the documentary fronted by Al Gore, *An Inconvenient Sequel: Truth to Power*, with its focus on the dangers of climate change, in summer 02017 (Andrews 2017).

Even in the field of economics, the concept of limits is now starting to be paid attention. An influential new term is that of the circular economy: an economy that is focused on addressing environmental issues at the design stage and aims to extend the life of products, components, and materials as much as possible (EMAF 2013; European Commission 2017). There is also room for such concepts to be applied in the ICT field.

9.1   A Longer-Term View of Sustainable ICT Over Time

A stage in human life on this planet is about to begin in which absolutely all of humankind's activities should be scrutinised by placing a strong and careful emphasis on sustainability. Some researchers (e.g. Zalasiewicz et al. 2010) have proposed the introduction of a new geological term to identify an epoch that began either with the start of the Industrial

Revolution or with the use of atomic energy. In geological terms, humanity is currently living in the Holocene epoch of the Quaternary period; instead, Zalasiewicz and colleagues' proposal is to use the term Anthropocene (Zalasiewicz et al. 2010). This relatively new term describes an era in which human activities have long-term, irreversible impacts on the ecosystem. To be positive, of course, this can also be seen as an epoch in which human beings can, and should, make a constructive difference to the sustainability of their planet.

A number of volumes have already started to address sustainability issues in the ICT field (Fairweather 2011; Whitehouse et al. 2011; Hilty 2008; Patrignani and Kavathatzopoulos 2012). Dedicated series of conferences are also being held on the topic (see, e.g. ICT4S 2014). These books and conferences have investigated subjects like Green information technology (Green IT), the meaning of social accountability and sustainability in the information society, the connections between ICT and environmental issues, and the long-term sustainability of ICT itself.

Hence, a long-term view of ICT needs to be introduced, that can be used wisely to decrease the environmental impact of humankind's ICT-related activities. For this reason, the work of Daniel Spreng (1993) is of special importance. It forms an area for our ultimate consideration, before finishing the book with a final reflection on human beings' search for long-term human happiness.

9.2   Spreng's Triangle

It is useful to search for methods that can help to expand human beings' view of their environment from one based on short-term intervals towards a longer-term view that would be more in harmony with the planet. A useful model for this purpose is the one proposed by Spreng in 1993. Spreng wrote that to do any job requires amounts of energy, time, and information. Reducing energy input is achieved by increasing the time and/or information input one spends on the job. An equilateral triangle is the result (Spreng 2013:6).

This interesting model was first proposed in 01993: this is a triangle in which energy, time, and information are three dimensions that are strongly correlated. In this context, an adapted version of the original model is used to show the evolution of the different uses of the three variables through different stages of human history (see Fig. 9.1).

Energy, Time, and Information

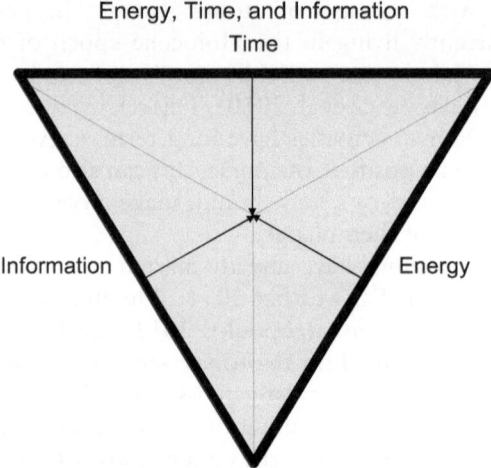

Fig. 9.1 Mutual substitutability of energy, time, and information (Authors' adaptation from Spreng 1993)

Each of the three dimensions of energy, time, and information is located along the three sides of an upturned triangle with three coordinates: for each point inside the triangle, energy is represented by the distance of the point from the side of the triangle, the time by the distance from the side related to time, and the information by the distance from the information-related side. The scales are purely qualitative.

Energy, time, and information correspond to the energy needed by any human activity, the time spent on it, and the information available or used. The point at the very centre of the triangle uses the same amount of all the three variables of energy, time, and information. If that point is shifted towards the energy side, while the amount of energy needed to perform the activity is diminished, the amount of time and information involved is increased (Spreng 1993).

A second adaptation of the model shows the evolution of the different uses of these three variables—energy, time, and information—through different stages of human history (see Fig. 9.2). It is there that ICT really comes into play.

Spreng's model (or triangle) clearly raised a reflection about time, speed, and qualitative approaches when, in the mid-01990s, he revived some earlier sustainability debates. While Spreng used his model to

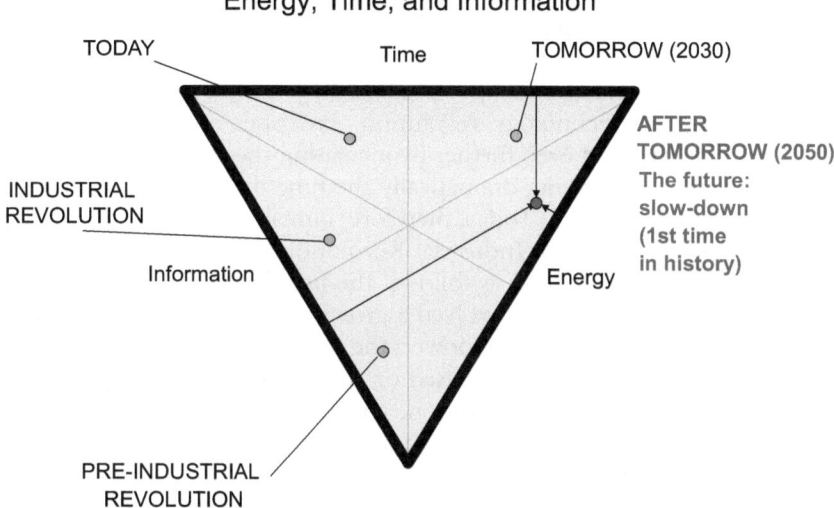

Fig. 9.2   Energy, time, and information (Authors' adaptation from Spreng 1993)

illustrate the position of activities and markets, today it can function as a means of facilitating a wider reflection. This whole debate can be extended to assess the role that ICT can have in human development, and ICT should be encouraged to play in the future.

This second view of the model permits an overview of a simple path of human history on planet Earth. It is useful to use this diagram to locate the several eras of human beings' technological and industrial advancement. Each stage is referred to by a number: stages 1–5:

1. Stage 1. In the pre-industrial stage (often considered to be before 01784), human processes and activities were very slow and hence consumed very low levels of energy and fossil fuels, whereas—in contrast—the amount of time needed for human activities was high. See the bottom left point of the triangle.

2. Stage 2. The acceleration of work processes and fossil consumption started with the first Industrial Revolution (01784 or thereabouts). In the Industrial Revolution, processes were accelerated. The increasing amount of energy used enabled people to speed up their activities by decreasing the amount of time used. The use of

information was, however, still low. See towards the top left-hand side of the triangle.

3. Stage 3. At the very top of the upturned triangle, along the middle of its 'base', humankind had reached 01950—the beginning of the information technology revolution. Processes and activities have been accelerated even further by increasing the amount of information used, decreasing dramatically the time dimension, and enlarging the energy used. Today, therefore, humankind is in 02017 in the middle of the fourth Industrial Revolution. (The first three revolutions can be described as follows: the first revolution—which took place around 01784—involved a growth in mechanical production, based on water and steam power; the second revolution, which took place around 01870, was based on mass production and the use of electricity; the third revolution, which took place around 01969, was based on electronics and an initial use of ICT for automation.) The fourth revolution is now ongoing. It is based on the use of 'cyberphysical' systems, extreme automation, robots, sensors that produce big data, artificial intelligence, and cloud computing. ICT is at its core (Hermann et al. 2016).

4. Stage 4. In future, an increasingly wise use of ICT could mean concentrating on the minimisation of environmental impacts and focusing less on speed. Looking ahead to 'tomorrow' (e.g. say 2030), one could forecast that a wiser use of information will enable society to make a shift towards sustainability. There could be a concerted reduction in energy consumed, which would be enabled by a definite increase in the amount of information use (e.g. through the energy saved by dematerialisation processes or by the use of smart grids for power generation and distribution). See towards the top right-hand side of the triangle.

5. Stage 5. Going beyond 'tomorrow' (say 2050), humanity will reach the right side of the triangle. It will be crucial to estimate carefully the amount of information to be used over the next 30 and more years, since ICT is reaching its limits in terms of its environmental impact. As a society, people will need to decrease the amount of energy that they use because the planet is reaching its sustainability limits. The dimension that is still available to human beings is time. By increasing time, human beings can keep the consumption of both information and energy low: to increase time for human activity means slowing down. This can be called After Tomorrow.

According to Spreng himself, ICT could contribute to long-term sustainability only if it is deployed in a more 'controlled and intelligent manner'. He therefore claimed that a wise use of ICT, applied discriminately, is needed because, in order 'to do things smarter, [be] wary of automation and higher speeds ... The debate on aiming at qualitative growth rather than quantitative growth held in the 1980s has almost been forgotten but needs to be revived' (Spreng 1993, 2013).

Of course, ICT can contribute to lowering energy consumption by increasing the use of information, but ICT could also, in general, accelerate many processes that could have potentially negative impacts on society and on the environment. This therefore calls for a vital new approach which involves searching for the benefits that ICT can bring and yet doing so in a reflective and considered way.

9.3   An Application of the Brakes with a Focus on Human Happiness

For the first time in human history, people need to rethink the speed of processes and find a new equilibrium with the environment.

While it is highly feasible that human beings will begin to use energy and information more wisely, it is important to accept that time cannot be compressed at any cost, speed is not a value in itself, and Slow Tech is needed. Slow Tech is a gentle application of brakes to the technological imperative (Patrignani and Whitehouse 2013: 388). It is an approach that can help people to understand what the implications of new technologies are and what changes humankind is opening itself up to in the future. It is thus important to make time for Slow Tech dialogue and discussion.

A reflection about time also includes serious thought and consideration about human beings' lives and the role of technologies. The key question is: can technologies (and ICT, in particular) support human beings' efforts in improving well-being, well-living, the search for happiness, and the containment of suffering?

What is the meaning of happiness, therefore, and what are its tenets? The leadership and governments of several countries have chosen to mention the word happiness in their foundational documents or current laws. Examples include 'the pursuit of happiness' in the United States' Declaration of Independence (USHistory 1776) or the definition of Gross National Happiness in the Constitution of Bhutan, which stipulates that:

'The State shall strive to promote those conditions that will enable the pursuit of Gross National Happiness' (Bhutan 2017). This latter, national indicator is in striking contrast with the more commonly internationally used measurement of Gross National Product, which represents the market value of the products and services of a country.

Querying the common meaning of happiness raises complex questions: Is happiness related to the amount of consumption? Is it related to the annual volume of goods, products, services consumed by a given population? Or is it related to the various qualities of food, water, air, health, instruction level, living conditions, and richness of relationships? Of course, any definition of happiness is profoundly connected with the meaning of life. But, in this case, there is a fundamental and evident difference: while, on the one side, products and market value are related to 'consumption' (the availability of quantity of materials or energy), on the other side, happiness is related to 'interactions' (and is, therefore, clearly, a more qualitative approach). If the first, quantitative, approach has brought humanity towards the verge of a deep social, economic, and environmental crisis, why not instead begin to experiment with visions of alternative, qualitative, dimensions of happiness? Researchers have concluded that a novel focus on Gross National Happiness might provide an appropriate, evidence-based framework for incorporating aspects of well-being related to health, social, environmental, economic, and cultural conditions into the measurement of societal advancement (Pennock and Ura 2011).

ICT could play a very important role in this search for the qualitative dimension of human happiness. Since information is an interaction—as can be seen at various points throughout this volume—it is by definition an intangible entity that could help to translate the concept of happiness into concrete projects in which good, clean, and fair information technologies are dedicated to improving the lot of human beings. As a result of such a reorientation and new focus, the quality of a wide range of essential products and services could all be enhanced and improved throughout the globe. Examples of such products and services include food (to be improved through precision agriculture and smart farms), water (monitored through applications related to the Internet of Things), air (in smart cities), health (supported via telecare centres), instruction levels (helped by holding virtual classes), living conditions (in zero-carbon homes), and the richness of human relationships (by the balancing of online and offline lives) (Bradley 2017), the reversion to a slower existence (Hendricks 2016), and generally switching off and powering down whether for reasons of ecology or mindfulness (Williams 2017). As captured by the content of a paper in a 02016 conference, ICT could become a mechanism for improving ecosystems and the social conditions of communities (Mata and Pont 2016).

It is evident that ultimately speed is becoming less and less important. Today, it is much more important to drive Slow Tech in the direction of the design and use of ICT applications that are searching for quality, not quantity. While Slow Food nudges human beings to 'drink less wine but drink better wine', Slow Tech could nudge the people of tomorrow towards a 'consume less but live better' lifestyle. The end result—it is hoped—would be a more responsible, sustainable, and ethical approach to ICT.

REFERENCES

Andrews, N. (2017, August 19/20). Films on Release. An Inconvenient Sequel: Truth to Power. *Financial Times*, p. 15.

Bhutan. (2017). *The Constitution of The Kingdom of Bhutan*. http://www.nationalcouncil.bt/assets/uploads/files/Constitution of Bhutan English.pdf. Accessed [9 October 2017].

Bradley, G. (2017). *The Good ICT Society. From Theory to Actions*. Abingdon/Oxon: Routledge Focus.

Di Paola, M., & Pellegrino, G. (Eds.). (2014). *Canned Heat: Ethics and Politics of Global Climate Change*. New Delhi: Routledge India.

Elliott, L. (2015). *Davos 2015: Climate Change Makes a Comeback*. www.theguardian.com/business/2015/jan/21/davos-2015-claimet-change-makes-comeback. Accessed [9 October 2017].

EMAF. (2013). *Circular Economy*. The Ellen MacArthur Foundation. Available at: ellenmacarthurfoundation.org. Accessed [9 October 2017].

European Commission. (2017). *Towards a Circular Economy*. https://ec.europa.eu/commission/priorities/jobs-growth-and-investment/towards-circular-economy_en. Accessed [9 October 2017].

Fairweather, N. B. (2011). Even Greener IT: Bringing Green Theory and 'Green It' Together, or Why Concern About Greenhouse Gasses Is Only a Starting Point. *Journal of Information, Communication and Ethics in Society, 9*(2), 68–82.

Franciscus. (2015). *Laudato—si'*. Rome: Libreria Editrice Vaticana. http://w2.vatican.va/content/francesco/en/encyclicals/documents/papa-francesco_20150524_enciclica-laudato-si.html. Accessed [9 October 2017].

Hendricks, J. (2016). *Be a Little Analog*. Biebesheim am Rhein: Thiele & Brandstätter Verlag GmbH.

Hermann, M., Pentek, T., & Otto, B. (2016, March 10). Design Principles for Industrie 4.0 Scenarios. *IEEE Explore*.

Hilty, L. M. (2008). *Information Technology and Sustainability. Essays on the Relationship Between ICT and Sustainable Development* Norderstedt: Books on Demand. http://www.bod.de. Accessed [9 October 2017].

ICT4S. (2014, August 24–27). *Proceedings of the 2nd International Conference on Information and Communication Technologies for Sustainability*, ICT4S 2014, KTH Royal Institute of Technology, Stockholm. Available at http://2014.ict4s.org. Accessed [9 October 2017].

Jonas, H. (1985). *The Imperative of Responsibility: In Search of an Ethics for the Technological Age* (H. Jonas & D. Herr, Trans.). Chicago: University of Chicago Press. Originally published as *Das Prinzip Verantwortung: Versuch einer Ethik für die technologische Zivilisation* [Frankfurt am Main: Insel Verlag, 1979].

Mata, F., & Pont, A. (2016, September 12–14). *ICT for Promoting Human Development and Protecting the Environment: 6th IFIP World Information Technology Forum*, WITFOR 2016, San José, Costa Rica, Proceedings, Vol. 481, IFIP Advances in Information and Communication Technology, Springer.

Meadows, D. H., Meadows, D. L., Randers, J., & Behrens, W. W., III. (1972). *The Limits to Growth*. Universe Books.

Patrignani, N., & Kavathatzopoulos, I. (2012, September 27–28). Is the Post-Turing ICT Sustainable? In *ICT Critical Infrastructures and Society*, Proceedings of 10th IFIP TC 9 International Conference on Human Choice and Computers, HCC10 2012, Amsterdam.

Patrignani, N., & Whitehouse, D. (2013). Slow Tech: Towards Good, Clean, and Fair ICT. In T. W. Bynum, W. Fleischman, A. Gerdes, G. M. Nielsen & S. Rogerson (Eds.), *The Possibilities of Ethical ICT*, Proceedings of ETHICOMP 2013—International Conference on the Social and Ethical Impacts of Information and Communication Technology (pp. 384–390), Print & Sign University of Southern Denmark, Kolding.

Pennock, M., & Ura, K. (2011). Gross National Happiness as a Framework for Health Impact Assessment. *Environmental Impact Assessment Review, 31*(2011), 61–65. Elsevier.

Skrimshire, S. (2010). *Future Ethics: Climate Change and Apocalyptic Imagination*. London: Continuum International Publishing Group.

Spreng, D. (1993, January). Possibility for Substitution Between Energy, Time and Information. *Energy Policy, 21*(1), 13–23.

Spreng, D. (2013, February 14–16). Interactions Between Energy, Information and Growth. In L. M. Hilty, B. Aebischer, G. Andersson & W. Lohmann (Eds.), *ICT4S 2013: Proceedings of the First International Conference on Information and Communication Technologies for Sustainability*, ETH Zurich.

UN. (2015, November 30–December 11). *UN Climate Change Conference*. Paris. http://www.cop21.gouv.fr/en. Accessed [9 October 2017].

USHistory. (1776). *The Declaration of Independence: Rough Draft.* http://www. ushistory.org/declaration/document/rough.html. Accessed [9 October 2017].

Vidal, J. (2015). Pope Francis's Edict on Climate Change Will Anger Deniers and US Churches. *The Guardian.* http://www.theguardian.com/world/2014/ dec/27/pope-francis-edict-climate-change-us-rightwing. Accessed [9 October 2017].

Whitehouse, D., Hilty, L., Patrignani, N., & Van Lieshout, M. (Eds.). (2011). *Social Accountability and Sustainability in the Information Society: Perspectives on Long-Term Responsibility,* special issue of Notizie di Politeia 27 (104).

Williams, A. (2017, August 26/27). Why a Dumb Phone Is a Smart Move. Tech World. Notes from a Digital Bunker. *FT.Com/Magazine.*

Zalasiewicz, J., Williams, M., Will Steffen, W., & Crutzen, P. (2010). The New World of the Anthropocene. *Environmental Science & Technology, 44*(7), 2228–2231. https://doi.org/10.1021/es903118j.

INDEX